600.21
Boy's

D0742722

Algrove Publishing Limited
1090 Morrison Drive
Ottawa, Ontario
Canada K2H 1C2

Canadian Cataloguing in Publication Data

Main entry under title:

The boy's book of trades : and the tools used in them

(Classic reprint series)
Reprinted from the edition first published: London : New York ;
G. Routledge and Sons, 1866.
Includes bibliographical references and index.
ISBN 0-921335-60-1

1. Industrial arts. 2. Tools. I. Series: Classic reprint series (Ottawa, Ont.)

HF5381.2.B69 1999 600 C98-901437-1

Printed in Canada
#31100

Publisher's Note

Originally published about 1865 as *The Boy's Book of Trades,* this book was obviously intended to provide a synopsis of a number of the more common trades so that potential apprentices might choose among them. As happened with so many books of that era, if an allied subject appeared to be interesting, it was tossed into the text somewhere. The portion of this book dealing with the manufacture of gas is a classic example of this tendency. The justification for including it was undoubtedly that it was related to the gas fitter's trade.

Despite this foible, the book draws together a great deal of information on trades that have long since vanished, making it a significant historical document.

Leonard G. Lee, Publisher
Ottawa
January, 1999

THE

BOY'S BOOK OF TRADES

AND THE

TOOLS USED IN THEM

COMPRISING

BRICKMAKER
MASON
BRICKLAYER
PLASTERER, &c
CARPENTER
PAINTER, &c
PLUMBER
MANUFACTURE OF GAS
IRONFOUNDER
BLACKSMITH
BRASSFOUNDER
GILDER

CABINET MAKER
FLOOR-CLOTH DITTO
PAPER STAINER
CALICO PRINTER
TINMAN
FARRIER
NEEDLE MAKER
HOTPRESSER
CUTLER
COTTON MANUFACTURER

TAILOR
TANNER
SHOEMAKER
SADDLER, &c
HATTER
MILLER
BAKER
SUGAR REFINER
DYER AND SCOURER
COPPERSMITH
GUN MAKER

WITH NUMEROUS ILLUSTRATIONS.

LONDON
GEORGE ROUTLEDGE AND SONS
BROADWAY, LUDGATE HILL
NEW YORK: 416 BROOME STREET

PREFACE.

Every human being born into the world will find that happiness mainly depends upon the work that he does and the manner in which it is done.

Those who imagine that the necessity for labour is only an evil must be either grossly ignorant or wilfully wicked.

Whoever wastes his life in idleness, either because he need not work in order to live, or because he will not live to work, will be a wretched creature, and at the close of a listless existence will regret the loss of precious gifts and the neglect of great opportunities.

Our daily work, however common or humble it may seem, is our daily duty, and by doing it well we may even make it a part of our daily worship.

For these reasons the choice of a trade is a most important event in every boy's life, and it is no less difficult than important, because when a boy has just left school he seldom knows much about the operations of any trade, and cannot be expected to express any preference for one more than another. Whether this book will be of any use in this respect, by directing attention to some of the principal industries of the country in which we live, must after all depend upon the tastes of each particular reader; but it may at all events claim to be useful in making known what are the operations necessary to some of our great manufactures, and in explaining the method of using the tools employed by those engaged in them.

A volume like the present can scarcely be said to have an Author, since much of its contents must necessarily be the work of the compiler, who condenses the valuable material supplied by other writers, and adapts it to his purpose. It is impossible that any great part of the *matter* can be original; and in some parts of the following pages the writer has found it altogether inexpedient to change even the *manner* of those who are regarded as the principal authorities on scientific subjects. He believes, however, that even where he has borrowed, he has borrowed with some discretion. Whether he has succeeded in explaining and simplifying the reader will be able to judge.

THE BRICKMAKER.

BRICKFIELD, SHED, KILN, &C.

It would be very difficult, and perhaps impossible, to discover at what time in the history of the world the art of brickmaking was first practised. In the earliest records of the human race the making of bricks is mentioned; this was part of the labour imposed upon the children of Israel, when they were in captivity in Egypt, and bricks of excellent quality are found in some of the most ancient buildings, the remains of which have been discovered. Though uncivilized nations, and even some which had made great progress in civilization, but lived in very warm or exceedingly cold climates, frequently built dwellings of wood, of wattles or strips of trees and branches covered with clay and lime, and of rough stone and earth; and though whole tribes lived and still live under tents, or in mere log huts and wigwams, or lodges made of the skins of animals, the manufacture of bricks formed of clay, and either burnt with fire or dried in the sun, is amongst the oldest of all known trades. In our own day it has arrived at such perfection and the varieties of bricks and tiles are so great in order to provide for the great diversity in buildings, that it is one of the most important branches of English industry.

We learn from the Bible that burnt bricks were used in building the Tower of Babel, and from early historians, as well as from recent discoveries, we know that they were also made for the walls of Babylon. The bricks of the ancient Egyptians were made of clay tempered with water and mixed with chopped straw, and afterwards dried in the sun, and the labour of the Israelites was made more severe by their being compelled to find straw for themselves. In Rome both burnt or kiln-dried bricks and those dried in the sun were employed, and though at a later date the art of brickmaking seems to have fallen into disuse, it was revived again in Italy after some hundreds of years. The trade seems to have been brought to England by the Romans, and many of the most ancient buildings in this country are made of very fine brickwork, though, till the reign of Elizabeth, only large mansions were so built, the common houses being formed of frame-works of timber filled in with coarse plaster supported by laths of wood.

There are few more interesting sights than a brick-field

in full work with its great sheds, its horses going slowly round and round in the mills, grinding the clay which has been dug out of the deep pits; its great stacks covered with hurdles and screens made of reeds and its immense kilns, so cleverly and evenly built, where the smoke rises lazily from the dull fires by which the drying or burning is completed.

Clay Mill.

The methods of brickmaking differ considerably in various parts of England, but that which we most commonly see in practice near London will very well represent them all, and it is this which will now be described. The earth used for making bricks is found after digging till the labourers reach the loamy soil lying just above that blue clay which is known as London clay; and this earth is known as strong clay, mild clay, and malm, and this earth requires preparation by mixing with them chalk and the dust of burnt ashes from the dust bins. These burnt ashes the brickmakers call "breeze."

The chalk mill and the clay mill are placed close together on large mounds, high enough to allow the "malm" (which is a mixture of chalk and clay ground to a thin paste) to run down to the brick earth. The chalk mill is a round trough where the chalk is ground by heavy wheels fitted with spikes on their tires or hoops, and turned by one or two horses. The trough is supplied with water from a pump, and the chalk, as it is ground, runs off by a wooden gutter into the clay mill, where it is again stirred and ground till it mixes with the clay; the mixture then runs through a grating and through other gutters to the brick earth, which has been placed in heaps to receive it.

Hack Barrow. Barrow.

When the earth is mixed in this way, it becomes brick-clay, and is taken in barrows up a sloping board, to the pug mill. The pug mill is a great tub, the top of which is larger than the bottom, and in the centre of it there revolves an upright iron shaft fitted with knives. These knives cut and break the clay as it passes through the mill, and they also force it downwards till it reaches the bottom, where it passes through a hole on to a machine called the Cuckhold, which is a sort of table containing a trough where the clay is cut into lumps ready for the moulder.

Pug Mill.

The moulder who shapes the clay into bricks uses moulding sand,—a peculiar sort of sand brought from the

Barrow for carrying baked Bricks

bed of the river, and spread out in the sun, where it is turned over and over till it is quite dry. It prevents the clay from shrinking, gives a harder surface to the bricks, and prevents them from sticking to the mould, or to each other; it also gives the London bricks their grey colour. The moulder stands at the moulding stool, which has a rim at each end to keep the moulding sand from falling off, and has a stockboard, which forms the bottom of the

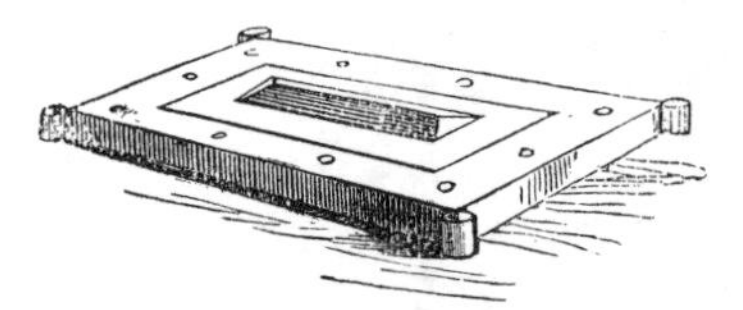

Kick and Stockboard.

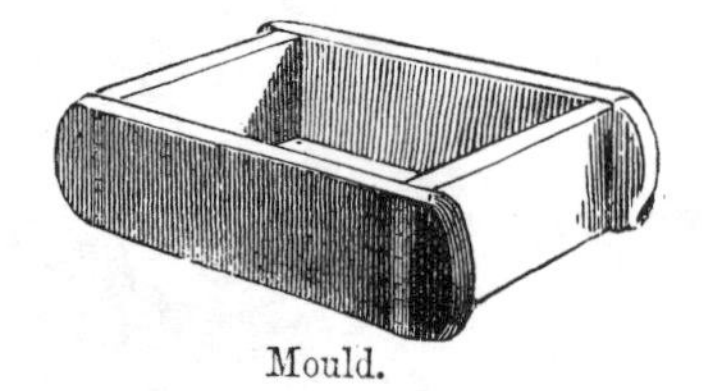

Mould.

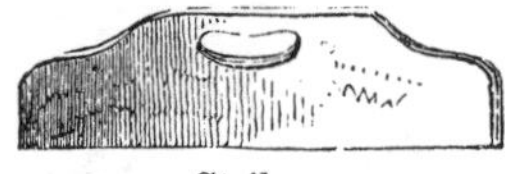

Board for Moulding.

Strike.

Brick-mould.

brick-mould, and a page, or two iron rods nailed at each end to wooden rails, used to slide the raw bricks from the moulder to the place from which the "taking-off boy" takes them to place on the "hack barrow," by which they are carried away. The moulder is served with the lumps of clay by the "clot moulder." The "brick-mould" is a kind of box without top or bottom, and the moulder dashes the tempered clay into the mould with sufficient force to make the clay completely fill it; after which the superfluous clay is removed from the surface of the mould with the strike. The

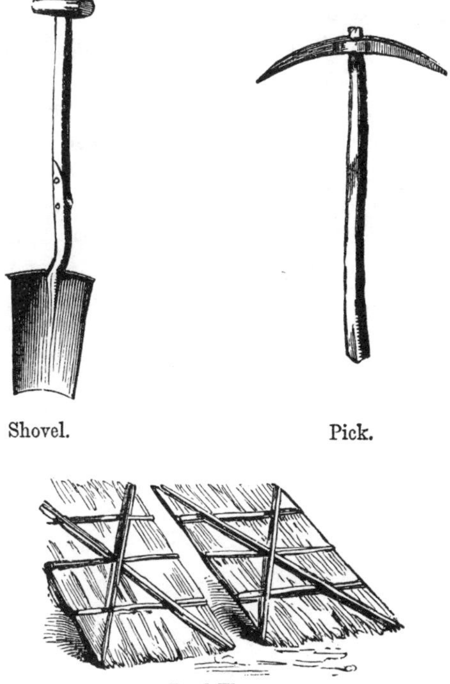

Shovel. Pick.

Reed Flats.

brick is then turned out on to a pallet or board, on which it is wheeled by the boy to the "hack ground," where the bricks are built up to dry in low walls called "hacks." The brick moulds are made of brass or iron, and often of wood.

Sometimes the bricks are dried on a floor under a shed, but often in the open air, where they are covered with straw, reed-flats, or canvas and tarpauline screens, to protect them from wet, frost, or excessive heat. The bricks are afterwards burnt either in "clamps" or in "kilns." In clamp burning the bricks are built up close together, and the bottom ones only are heated with burning breeze or cinders, the heat spreading to those at the top. A kiln is a sort of large chamber in which the bricks are loosely stacked with spaces between them for the heat to pass through, and they are baked by fires placed either in arched furnaces under the floors of the kiln, or in fire holes made in the side walls. The kilns are built of various shapes, and one of the principal arts in the trade of the Brickmaker is to construct them that the heat may be properly distributed, and the bricks equally and thoroughly baked.

THE MASON.

MASONS AT WORK.

HAVING already given a description of the way in which bricks are made, we come to the work of the Mason, whose duty it is to prepare the stone work used in building and for other purposes. In the mason's trade great skill is required, as well as some hard manual labour, since he has to cut the stones for arches, windows, columns, cornices, and porches, into various shapes; and to fit the separate pieces with perfect accuracy, that the Builder or the Architect may be able at once to set them in their proper places. The business of the monumental mason, who erects pedestals for statues, tombs, and ornamental structures in parks and gardens, is generally distinct from that of the builder or architectural mason, although many of the same tools are used in both trades.

The stone used by the mason is of various kinds, and is brought from different parts of the world; but our own country contains stone of nearly every sort which can be well employed in ordinary building. Granite comes principally from Scotland, though a smaller quantity is brought from Ireland; red and white sandstone is plentiful in Yorkshire, Lancashire, and Derbyshire, as well as in Scotland and Ireland; a sort of slate stone is found in Wales; and the most common building stone, which is called limestone, or free-stone, is brought from several counties in England, where it is constantly worked. The stone is generally found under the surface of the earth, and the places from which it is dug are called quarries; the business of quarrying being to extract from the ground, or from the sides of rocks, large masses of stone or marble.

When these lie directly under the ground, the earth at the top is removed, and the stone is afterwards separated into blocks and lifted out by machinery; but it is sometimes necessary to mine for the stone by making galleries underground, and leaving pillars to support the earth above them. In large quarries, the earth at top is first removed, and the first layer of stone, which is generally of a common sort, is broken or blasted with gunpowder, and afterwards taken away. The lower layers of stone are then divided by wedges driven into them, until they split in the required direction; the blocks are afterwards made of a regular square, by a tool called a *kevel*, pointed at one end and flat at the other, and are then lifted by cranes on to low waggons, upon which they are drawn away. It is the business of the mason to work these stones, which are to be used in building, into their required shape; but before the mason receives them the *stone cutter* hews and cuts the large blocks roughly into the form in which they are wanted; and when the block is to form top of a doorway, part of a cornice, or any other portion of a building where ornament is necessary, the *carver* executes these ornaments, and cuts the stone into a pattern of fruit, flowers, or figures.

Peck or Point.

Stone Axe.

When the stone is valuable it is sent from the quarry to the mason's yard, or to the building where it is to be used in large blocks, and there cut into slabs or thin pieces called "scantlings," of the required size, with a stone mason's saw (*see large cut*). This saw differs from those used in other trades because it has no teeth. It is a long thin plate of steel slightly jagged on the bottom edge, and fixed in a frame; and being drawn backwards and forwards in a horizontal position, cuts the stone by its own weight. To make this the easier, a heap of sharp sand is placed on a sloping board over the stone, and water trickling upon it from a barrel washes it into the cut made by the saw. In large establishments the sawing of the stone is often effected by steam machinery. Some of the freestones are so soft as easily to be cut with a toothed saw, worked backwards and forwards by two persons. The tools used by the mason are the peck or point for chipping the surface of the stone, the stone axe for breaking the irregular portions from the block, the iron mallet and beetle, for breaking pieces from the edge of the slab or driving in wedges.

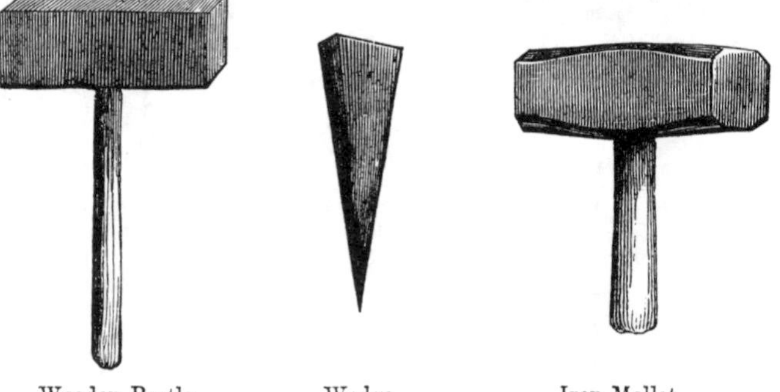

Wooden Beetle. Wedge. Iron Mallet.

The tools used for cutting stone are the mallet and chisels of various sizes. The mason's mallet differs from that used by other workmen, being of a sort of half pear shape, and with a short handle only just long enough to allow it to be firmly grasped in the hand. The rubber is used for smoothing the surface of the stone, after it has been worked by the tools;

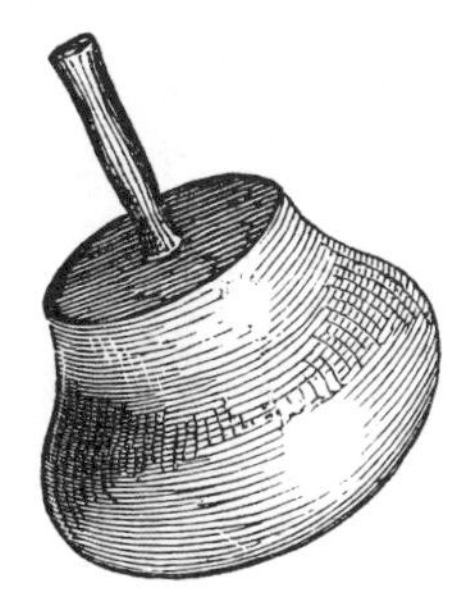

Mallet.

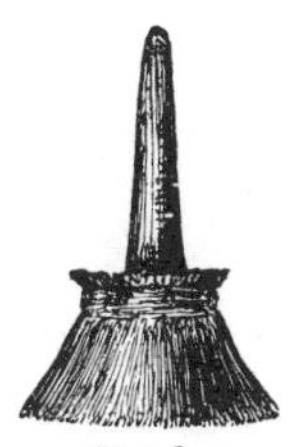

Brush.

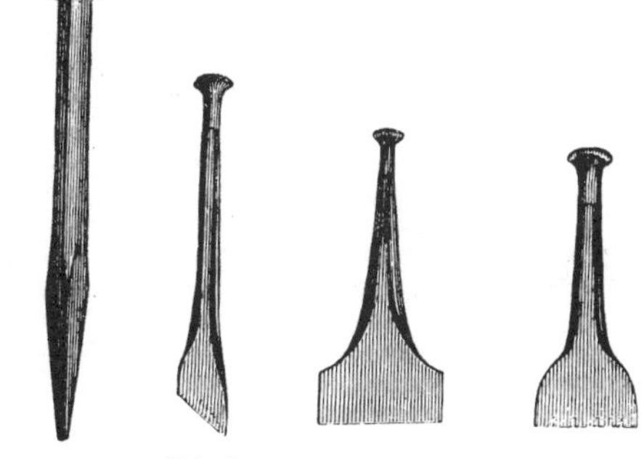

Chisels.

sand and water are placed on the stone, and the rubber is pushed backwards and forwards for the purpose of grinding the face of the slab to a smooth surface; another block of stone is sometimes applied

Rubber.

to the same purpose. In London the tools used to work the faces of the stone are—the *point*, a very small chisel only about a quarter of an inch broad at the cutting edge; the inch tool, which is a broader chisel; the *booster*, broader still; and the *broad tool*, which is three inches and a half wide: beside these, there are tools of the same kind for working mouldings and carvings.

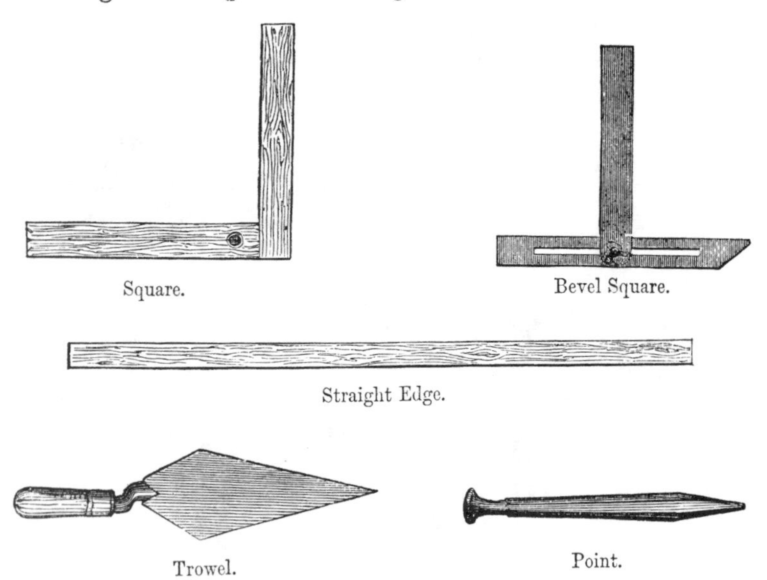

Square. Bevel Square.

Straight Edge.

Trowel. Point.

Besides these cutting tools, the mason uses a *banker* or bench, on which he places his stone for convenience of working, and *straight edges, squares, and bevels*, for marking the shapes into which the blocks are to be cut, and seeing that his edges and surfaces are even by trying them as the work proceeds.

The bevel square is a square the *stock*, or lower part, of which is moveable, so that it may be set to any angle or level as required. Sometimes a pattern called a *templet* is used for cutting a block to any particular shape, and when the work is moulded, the templet is called a *mould*. Moulds are commonly made of sheet zinc, carefully cut to the profile of the mouldings with shears and files.

It often happens that the mason has not only to prepare the stone, but to set it in its place in the building, and this is properly part of his work. He then uses the trowel, for applying the composition for cementing the stones together; lines and pins to show whether his edges are straight and square, the square and level for a similar purpose; and various rules for adjusting the stone faces of upright walls.

THE BRICKLAYER.

THE BRICKLAYER.

THE Bricklayer has so much to do with the erection of buildings that the Master Bricklayer is generally a Builder; that is, he understands not only brickwork and the building of walls, but also the other trades necessary for completing a house, and can superintend the Mason's, Carpenters', and Plumbers' work.

Of course, the first consideration in building a house is

the preparation of the foundations, which are formed in various ways, according to the nature of the soil on which they are laid; unless the ground itself is firm enough to receive the walls, sometimes thick layers of concrete (a sort of mortar) is used, sometimes layers of planking are put down, or even cross beams of timber, and in some cases, where the earth is very loose and damp, timber piles are driven into it on which to lay the foundation. The foundation once laid, the vaults or cellars are built either in the ordinary way, or in a series of arches of various forms; then follow the abutments, the wing-walls, the main-walls, with iron "Bressumers" for supporting those parts of the walls which are above large openings like great doors or shop fronts. The partitions, or interior walls, may be either solid brick or stone, or may be constructed entirely of timber, or they may be frames of timber filled in with masonry or brickwork. Then come the floors, the roofs and roof-coverings, and finally the ceilings and the doors and windows. The materials used in building are principally timber, stone, slate, bricks, tiles, mortar, lime, cement, iron, glass, lead, zinc, colours and varnishes. Those with which the working bricklayer has most to do are bricks, slates, tiles, stonework, cement, and mortar, for these are principally used in making walls and roofs, which is the greater part of his trade. The stone and slate come from the quarries, and we have already seen how bricks are made. Mortar is made in the following way.

1st. The soft chalky stone known as limestone, is calcined or burnt, by exposure to strong heat in a lime kiln; the heat drives off a gas which is contained in it (called carbonic acid gas), and leaves it in a state in which it is known as quick-lime.

2d. The quick-lime is "slaked," by pouring water upon it, when it swells and becomes very hot, afterwards falling into a fine powder.

3d. This powder is mixed into a rather stiff paste, more water is added, and when a certain quantity of sand is added becomes *mortar*, and may be used for cementing bricks together.

Concrete is made by mixing gravel, sand, and ground unslaked lime together with water; it is used for foundations, and filling in apertures requiring strength and firmness.

Iron is used by the builder in two different states, as cast iron and wrought iron; the *girders* for supporting roofs and walls are mostly cast in moulds *(See Iron Founder)*, though both these and other parts of the ironwork used in building, are frequently of iron wrought by hand.

Lead is used by the mason for securing and coating the iron clamps which hold the blocks of stone together; it is also used in the plumbers' work of a house *(See Plumber)*.

Zinc is used in the manufacture of gutters, pipes, and portions of roofs.

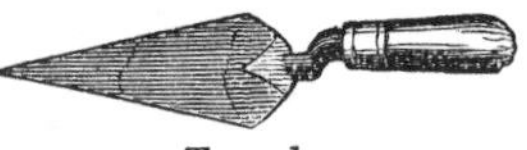

Trowel.

The excavator having dug out the space where the foundation of the house is to be laid, the work of the bricklayer begins, and his tools are the *trowel*, to take up and spread the mortar, and cut bricks by a sharp blow to the requisite length; then there are the *brick axe* or hammer, for shaping the bricks to a level; the *tin saw* for making an incision on bricks to be cut with the axe; the rubbing stone on which to rub the bricks smooth in the parts where they have been cut, and the *mortar rake and shovel,* for mixing the mortar and cement.

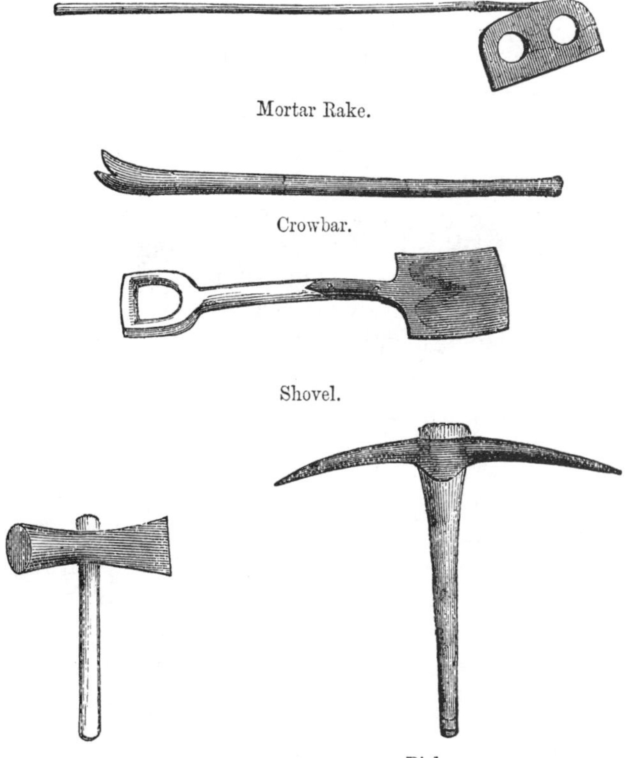

Mortar Rake.

Crowbar.

Shovel.

Brick Hammer.

Pickaxe.

The crowbar and pick-axe are the tools used for demolisning old brickwork or clearing out rubbish; the sieve *(see large cut)* for sifting the lime of which the mortar is made, and the rammer for hardening the ground to render the foundation firm. The *raker* is used for raking out the mortar from the joints of old brickwork which requires re-pointing, or the joints refilled with mortar.

To "set out" the work, which means to measure the spaces, and to keep the lines, curves, and angles, straight and true, the bricklayer uses the *square*, the *level*, the *plumb-rule*; the square shows whether the proper angle has been preserved; the plumb-rule is an upright rule, with a string at one end, to which is attached a leaden ball. If the work is straight on its perpendicular lines, and the plumb-rule be applied to it, this leaden ball will hang exactly in the centre, and swing through a hole in the rule, while, if the work be crooked, it will swing to one side. The level is a rule on the same principle, but for testing horizontal lines, such as a cornice on the top of a wall. The good bricklayer will frequently test his work by these tools, and will also use a line stretched to two pins, to guide him as he builds up his courses of bricks.

The bricks and mortar are supplied to him by a labourer, who carries them in a *hod*. The labourer also makes the mortar, and builds up and takes down the *scaffolding (see large cut)*. The scaffolding, or that frame of poles and planks erected in front of the building as it is in progress, is constructed of *standards*, *ledgers*, and *putlogs*.

The standards are poles made of fir trees, from forty to fifty feet long, and six or seven inches thick at the butt ends, which are firmly fixed in the ground. When one pole is not long enough, two are lashed together, the rope lashings being tightened by wedges driven in between the coils in a peculiar way. The ledgers are horizontal poles placed parallel to the walls and lashed to the standards; these support the putlogs, or cross pieces, which are about six feet long, one end of them resting in the wall, the other on the

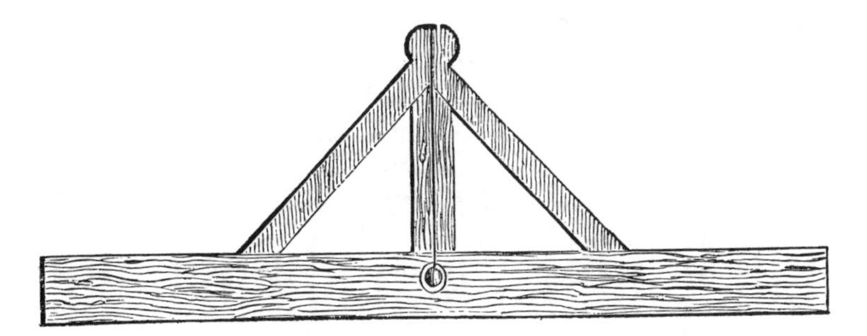

Level.

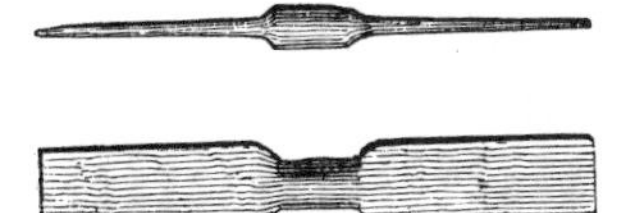

Cutting Chisel.

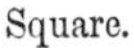

Square.

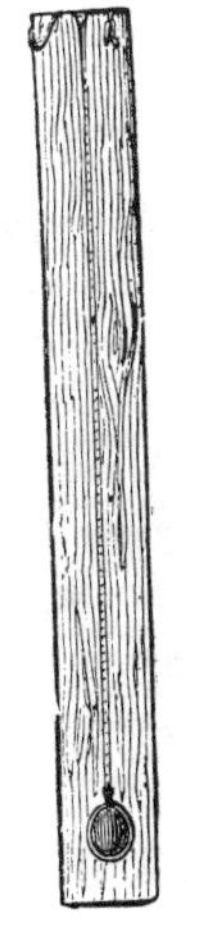

Line and Plumb.

Hod.

ledgers. On these putlogs are fastened the scaffold boards, which are stout boards with pieces of iron hoop placed round the ends, to keep them from splitting.

A bricklayer and his labourer will lay about a thousand bricks, or two cubic yards of brickwork, in a day.

The tools used for *tiling*, or placing the tiles on a roof, are the lathing hammer, the iron lathing staff to clinch the nails, the trowel, which is longer and narrower than that used for brickwork, the *bosse*, for holding mortar and tiles, with an iron hook to hang it to the laths or to a ladder, and the *striker* a piece of lath about ten inches long for clearing off the superfluous mortar at the feet of the tiles.

THE PLASTERER AND WHITEWASHER.

WHITEWASHER AND PLASTERER.

When the walls, or what is called the carcase of the house, have been built, the roof made, the inner walls and partitions set up, and the joists and woodwork of the floors laid down, the work of the Plasterer begins. He covers the brickwork and bare timbers of walls, ceilings, and partitions with plaster, to prepare them for painting or papering: he also forms the cornices for ceilings, and the mouldings and decorations, which are usually made in plaster or cement.

The materials which the Plasterer uses for these purposes are:—1st. *Coarse stuff*, or a paste made with lime, much in the same way as common mortar, and afterwards mixed with hair, which is obtained from the tanner's yard, after it has been removed from the skins, which are there made into leather. This hair is raked together, and mixed with the mortar, with the *hair hook*, and a sort of three pronged rake called the *drag*.

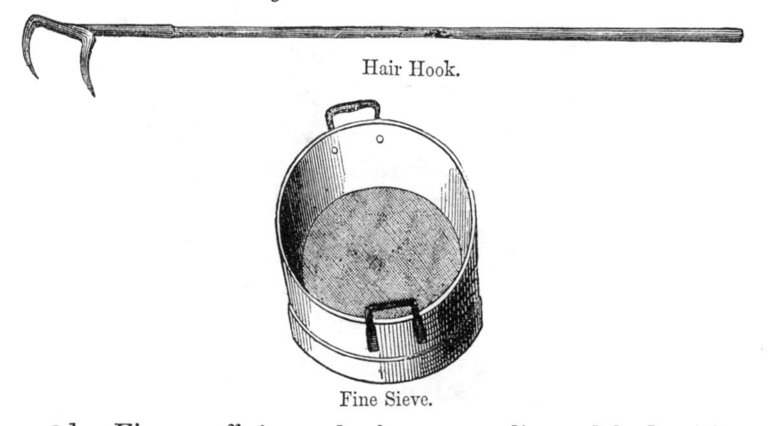

Hair Hook.

Fine Sieve.

2d. *Fine stuff* is made from pure lime, slaked with a small quantity of water, after which enough water is added to bring it to a state in which it resembles cream; it is then left to settle, the superfluous water is poured off, and the mixture is left in a tub, till still more of the *water* has evaporated, and it is thick enough to use. This stuff is often used for ceilings, and then a small quantity of white hair is mixed with it, to help to make it firmer and more binding.

For these finer kinds of plaster it is necessary to use the *fine sieve*, in order to sift the lime and other ingredients, that only the portion which has been reduced to powder may be retained.

3d. *Stucco*, which is made by mixing fine stuff with cleaned-washed sand. Stucco is used for house fronts, or other finishing work, which is intended to be painted.

4th. *Gauged stuff* is used for forming cornices, which run round a ceiling, and for mouldings; it is made by mixing fine stuff with plaster of Paris, which is a fine white powder easily made into a paste, and drying very quickly.

These are only the ordinary materials used by the Plasterer for his work, but there are a great number of cements, which are also applied in the course of his trade; such as Roman and Portland cements, and mixtures made and sold for special purposes of decoration, and the manufacture of ornaments.

These ornaments, such as centre pieces for ceilings, flowers drooping from cornices, bosses or groups on walls, &c. are first modelled in clay, and are afterwards cast in plaster of Paris, placed in moulds made of wax or plaster. In this trade, as in that of the picture frame maker, ornaments are frequently made of *papier maché*, or the pulp of paper (literally *smashed paper*), which is a very light, hard, and durable substance for the purpose. The moulds for cornices are made of sheet copper, and are fixed in a wooden frame.

The various tools used by the Plasterer are shown in the engravings, and the manner in which some of them are used may be seen by the large cut at the commencement of this description of the Plasterer's trade. The peculiar *hammer*, with one edge like an axe, is used for breaking down old plaster, and clearing away the mortar from walls and ceilings previous to plastering them afresh; the *hawk* is a flat board with a handle in the centre, used for holding the plaster or cement, which is being laid on with the trowel. The *gauging trowel* is the long narrow trowel, used

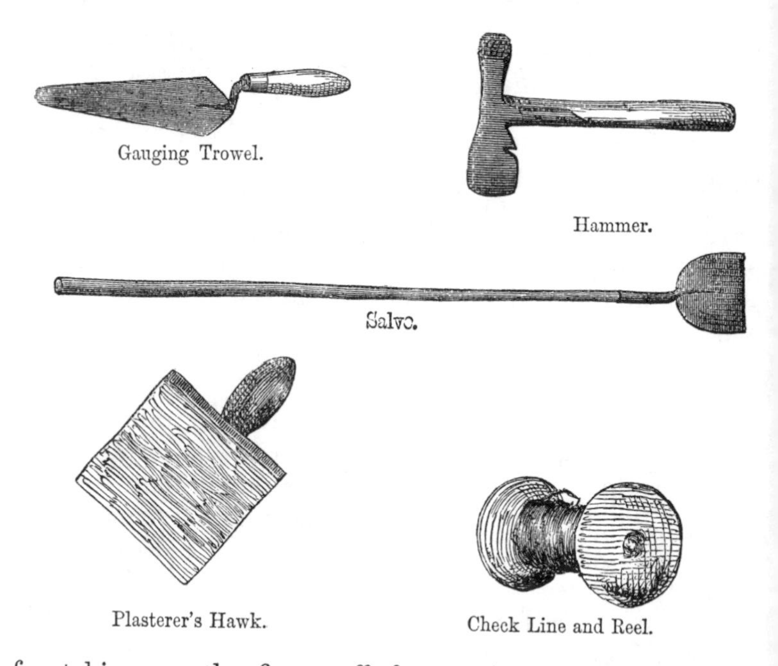

Gauging Trowel.

Hammer.

Salvo.

Plasterer's Hawk.

Check Line and Reel.

for taking up the fine stuff for cornices and mouldings; these trowels are of various lengths, from three to seven inches. The *salve* is a sort of small spade, on which the plasterer's boy lifts the mortar or cement, and places it on the *hawk*, which the workman holds in his hand; as the *salve* has a long handle, the plaster can be conveyed to the hawk even when the man is at work on a ladder. The square is similar to that used by the bricklayer, and shows that the corners of the work are straight and even; the *compasses* are used for measuring distances; the *check line*

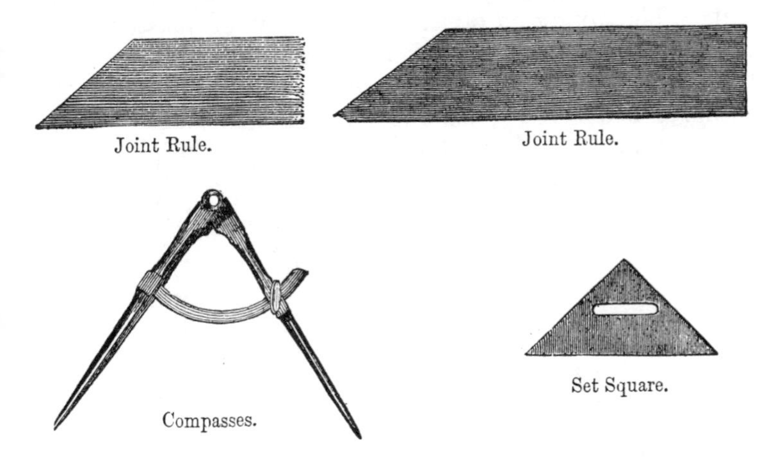

Joint Rule.

Joint Rule.

Compasses.

Set Square.

for marking out the spaces of the work to be done, and the *joint rules* for measuring the parts where different portions of the work come together, either at the corners, or in making the mouldings. The *set square* is also used, for showing that the surface or the line of the work is straight and even.

When the Plasterer has to cover a ceiling or a partition, he commences by lathing. This is nailing laths over the whole space which is to receive the plaster. Laths are long narrow strips of either oak or fir wood, of various thicknesses; the thicker being used for ceilings, where they have to bear a greater strain than in upright walls.

The next operation is *pricking up*, or placing the first coat of *coarse stuff* upon the laths; this is called pricking up, because when the plaster is laid, its whole surface is pricked and scratched with the end of a lath, that it may be rough enough for the next coat. The laying on of this second coat of plaster is called floating, and is performed in the following way. The surface is surrounded with narrow strips of plastering, called *screeds*, held fast by lines of nails, and these are made perfectly level, by means of the plumb

Square.

Modelling Tools.

Brush.

rule *(see Builder's tools)*, and the use of the *hand float*. The spaces within these lines of plaster work are then filled with coarse stuff, till the whole forms a flat surface, which is made perfectly level, or "floated" with the *floating rule*. Other screeds are then formed and filled up in the same way, until the whole ceiling, or wall, forms one flat surface. The operations are the same for ceilings and walls, except that the plumb rule is used for adjusting the level of walls, and the *level* for that of ceilings.

Floating Rule.

After the work has been brought to an even surface with the floating rule, the Plasterer goes over it again with the *hand float*, using a little soft stuff to make good any deficiencies.

When the floating is about half dry, the setting or finishing coat of fine stuff is laid on, and is first wetted with a

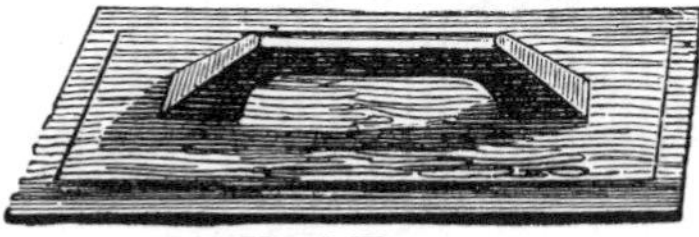

Hand Float.

brush, and then worked over with a smoothing tool until a fine surface is obtained.

Stucco is laid on with the largest trowel, and worked over with the *hand float*—being at the same time sprinkled with water—until it becomes hard and solid; after which it is rubbed over with a dry brush. The water has the effect of hardening the face of the stucco; which, after several sprinklings and trowellings, becomes very hard, and as smooth as glass.

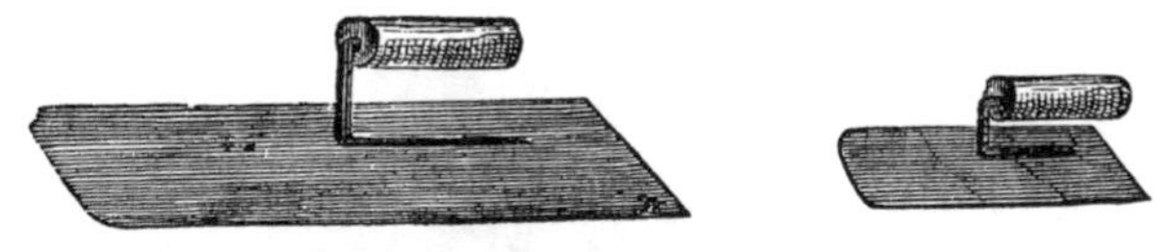

Setting Trowels.

The commonest kind of Plasterer's work is laying on one coat of stuff: when this is done on brickwork it is called *rendering*, and when on laths it is called *laying*. When there is a second coat it is called *render set*, or *lath-lay and set*: and when it is three-coat work it is called *render*, *float, and set,* or, *lath-lay, float, and set;* this is done in ceilings and partitions with fine stuff mixed with hair; or, when the walls are to be papered, with fine stuff and sand.

Rough stucco is used for finishing staircases and passages in imitation of stone. It is mixed with a great deal of coarse sand, and is not smoothed, the *hand float* being covered with a piece of felt, so that when it is applied to the stucco the grit of the sand sticks to it and is drawn to the surface, giving the plaster the appearance of rough stone.

Rough casting is used for outside walls, and is done by throwing a layer of gravel, mixed with lime and water, over the second coat of plaster while it is quite wet. In some counties of England, and especially near Nottingham, the plasterers use reeds instead of laths; and even floors are often made by laying down a quantity of coarse stuff upon a foundation of reeds.

These floors are almost as hard as stone, and possess the good quality of being almost always fireproof.

THE CARPENTER.

THE CARPENTER'S SHOP.

An account of the tools and implements used for working in wood would scarcely be complete without some remarks upon wood itself, and you can have no better information on this subject than that which has been written by a gentleman* who is thoroughly acquainted with the different kinds of timber, and with all the materials used in building.

* Mr. E. Dobson, Assoc. Inst. C. E.

If we examine a transverse section of the stem of a tree, we perceive it to consist of three distinct parts; the *bark*, the *wood*, and the *pith*. The wood appears disposed in rings round the pith, the outer rings being softer, and containing more sap, than those immediately round the pith, which form what is called the *heart-wood*.

These rings are also traversed by rays extending from the centre of the stem to the bark, called *medullary rays*.

The whole structure of a tree consists of minute vessels and cells, the former conveying the sap through the wood in its ascent, and through the bark to the leaves in its descent; and the latter performing the functions of secretion and nutrition during the life of the tree. The solid parts of a tree consist almost entirely of the fibrous parts composing the sides of the vessels and cells.

By numerous experiments it has been ascertained that the sap begins to ascend in the spring of the year, through the minute vessels in the wood, and descends through the bark to the leaves, and, after passing through them, is deposited in an altered state between the bark and the last year's wood, forming a new layer of bark and sap-wood, the old bark being pushed forward.

As the annual layers increase in number, the sap-wood ceases to perform its original functions; the fluid parts are evaporated or absorbed by the new wood, and, the sides of the vessels being pressed together by the growth of the latter, the sap-wood becomes heart-wood or perfect wood, and until this change takes place it is unfit for the purposes of the builder.

The vessels in each layer of wood are largest on the side nearest the centre of the stem, and smallest at the outside. This arises from the first being formed in the spring, when vegetation is most active. The oblong cells which surround the vessels are filled with fluids in the early growth; but, as the tree increases in size, these become evaporated and absorbed; and the cells become partly filled with depositions of woody matter and indurated secretions, depending on the nature of the soil, and affecting the quality of the timber. Thus Honduras mahogany is full of black specks, while the Spanish is full of minute white particles, giving the wood the appearance of having been rubbed over with chalk.

The best time for felling trees is either in mid-winter, when the sap has ceased to flow, or in midsummer, when the sap is temporarily expended in the production of leaves. An excellent plan is, to bark the timber in the spring and fell it in winter, by which means the sap-wood is dried up and hardened; but as the bark of most trees is valueless, the oak tree (whose bark is used in tanning) is almost the only one that will pay for being thus treated.

The seasoning of timber consists in the extraction or evaporation of the fluid parts, which are liable to decomposition on the cessation of the growth of the tree. This is usually effected by steeping the green timber in water, to dilute and wash out the sap as much as possible, and then drying it thoroughly by exposure to the air in an airy situation. The time required to season timber thoroughly in this manner will of course much depend on the sizes of the pieces to be seasoned; but for general purposes of carpentry, two years is the least that can be allowed, and, in seasoning timber for the use of the Joiner, a much longer time is usually required.

Properly seasoned timber, placed in a dry situation with

a free circulation of air round it, is very durable, and has been known to last for several hundred years without apparent deterioration. This is not, however, the case when exposed to moisture, which is always more or less prejudicial to its durability.

When timber is constantly under water, the action of the water dissolves a portion of its substance, which is made apparent by its becoming covered with a coat of slime. If it be exposed to alternations of dryness and moisture, as in the case of piles in tidal waters, the dissolved parts being continually removed by evaporation and the action of the water, new surfaces are exposed, and the wood rapidly decays.

Where timber is exposed to heat and moisture, the albumen or gelatinous matter in the sap-wood speedily putrefies and decomposes, causing what is called rot. The rot in timber is commonly divided into two kinds, the *wet* and the *dry*, but the chief difference between them is, that where the timber is exposed to the air, the gaseous products are freely evaporated; whilst, in a confined situation, they combine in a new form, viz. the dry-rot fungus, which, deriving its nourishment from the decaying timber, often grows to a length of many feet, spreading in every direction, and insinuating its delicate fibres even through the joints of brick walls.

In addition to the sources of decay above mentioned, timber placed in sea water is very liable to be completely destroyed by the perforations of the worm, unless protected by copper sheathing, the expense of which causes it to be seldom used for this purpose.

In modern houses the labours of the Builder, the Mason, and the Plasterer, would be of little use unless they were accompanied by that of the Carpenter, since a very large proportion of every building consists of the woodwork of which its interior structure is greatly composed.

As it is one of the most useful, so the Carpenter's may be considered the most ancient of trades, for nearly all other handicrafts require the preparation or manufacture of the materials, but the Carpenter originally found his materials in the forest, and at once set to work to construct various articles from the trunks and stems of the trees best suited for the purpose. We can only imagine one trade older than that of the Carpenter, and that is the Tool Maker, and as the earliest tools, or at at all events some portion of them, were probably made of hard wood, the Tool Maker may in some sense be said to have been a Carpenter also.

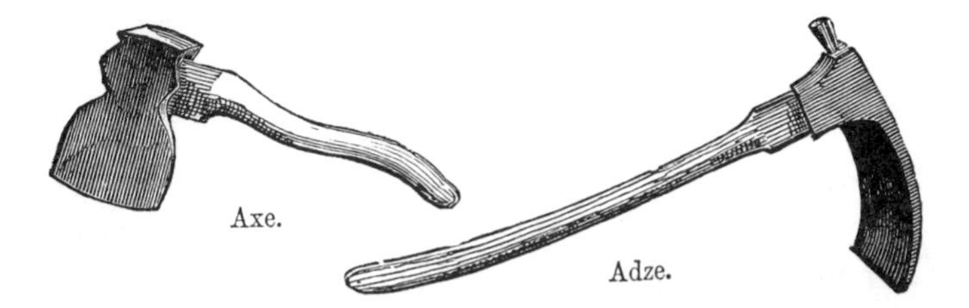

Axe. Adze.

Strictly speaking, the business of the Carpenter is only with the larger portions of buildings and the rough timber frameworks which support them, and his principal tools are the *axe* and the *adze*, for chopping and roughly smoothing timbers; the *saw*, for sawing beams and planks; the *chisel*, for making mortis holes for joining beams together, and for cutting and paring wood; the *chalk line*, a line rubbed with chalk, and used to make a straight line upon a board or beam, to mark the direction in which it is to be sawn; the *plumb rule*, already described amongst the Builder's

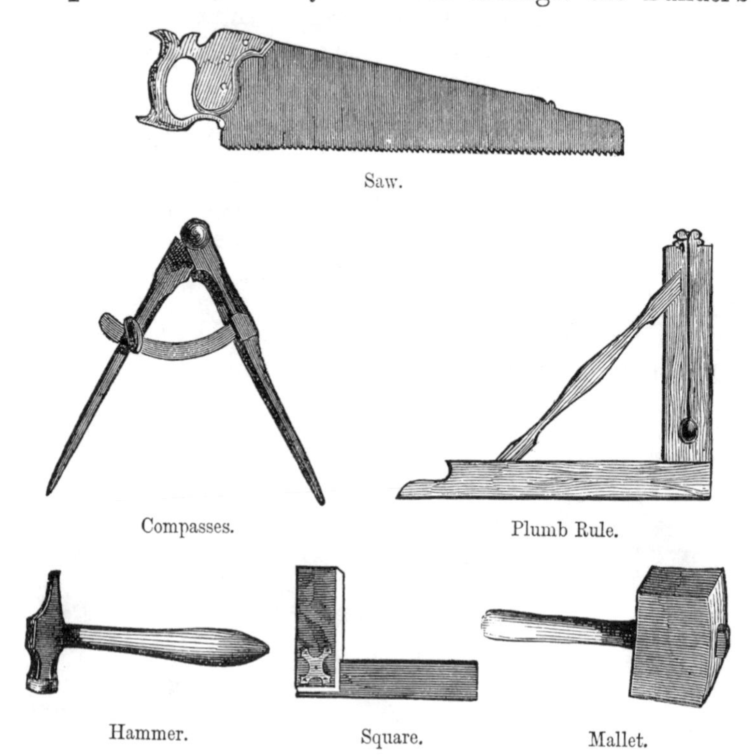

Saw.

Compasses. Plumb Rule.

Hammer. Square. Mallet.

tools; the *level*; the *square*; the *compasses*; all of which have been described in previous trades; the *hammer*; the *mallet*, and various sorts of nails. The other tools represented in the engravings belong more properly to the

Joiner, but as the trades of the Carpenter and the Joiner are almost always united, we will speak of all the tools as belonging to one business. Carpenter's work, then, consists of the framing roofs, partitions, and floors, in making the various joints used in beams, ties, rafters, and joists for supporting floors, and the proper way of supporting buildings by posts and girders. The Carpenter requires to be strong and active, that he may properly handle the heavy timbers on which he has to work; he should have a knowledge of the science of mechanics, that he may be able to provide for the strains and thrusts to which the different parts of his work are exposed, and supply the proper means of resisting them; and he should also be able to understand how to make what are called "*working drawings*," that he may "set out," or properly draw a plan of the work he has to do, from the designs of the Architect.

The Carpenter being concerned with the portions of a building which are made of timber, you will be better able to understand his trade by a short description of what these are; and we will then speak a little of the *Joiner*, whose trade is generally confounded with that of the Carpenter.

First: Partitions, or inner divisions of a building, may be either of brick, stone, or wood; and, in the latter case, they are generally "framed," or supported in a more solid framework, which should form a portion of the main building, that is, of the outer wall; and should be quite independent of the floors, which should not support, but be supported by them.

Second: Flooring is formed by joists or strong beams of timber reaching from wall to wall, where they rest upon other beams, called wall plates, which are built into the walls themselves. The floor boards are nailed over the upper edges of the joists, and the lower edges receive the laths and plaster, which form the ceilings of the rooms beneath. Large buildings are sometimes fitted with double framed floors, with two sets of joists, and building joists resting on girders; and in superior houses, the *wall plates* are often supported by "corbels," or, portions of the timber projecting from the inside of the wall, which prevents the necessity of opening the wall to admit the ends of the joists.

Third: Roofing consists of the roof covering, which is laid upon rafters or slender beams, which are supported by stronger horizontal beams called *purlins*; and these, again, rest on upright *trusses*, or strong frames of timber, placed on the walls at regular distances from each other. Upon the strength and firmness of these trusses, and the skill of the carpenter's work, depends the entire safety of the roof.

Large roofs are supported by cross beams, called *collars*, or *tie beams*; and they are further strengthened by an upright pillar in the centre, called a *King post*, from which slanting beams, called "struts," support the rafters.

We have now only spoken of common roofs, but there are many roofs of open timber-work in churches and other public buildings, which are wonderful specimens of the skill of the Carpenter and the Joiner. One of these, perhaps the grandest as well as the most ancient, is that of Westminster Hall; but there is one also in the Great Hall, at Hampton Court Palace; and others may be seen in Churches and Halls in various parts of England.

The Joiner, as his name implies, frames and joins together the wooden finishings and decorations of buildings, such as floors, staircases, skirtings, door and window frames, sashes, or, the sliding parts of windows that contain the glass, shutters, doors, chimney-pieces, &c. This work requires much greater nicety and finish than that of the Carpenter, and is brought to a smooth surface with the plane wherever it is likely to be seen, while the carpenter's work is left rough as it comes from the saw.

The principal cutting tools used by the Joiner are, saws, planes, and chisels.

The *saws* are of various sizes, and are called rippers, half rippers, hand saws, and panel saws, according to their shape, and the number of teeth to each inch.

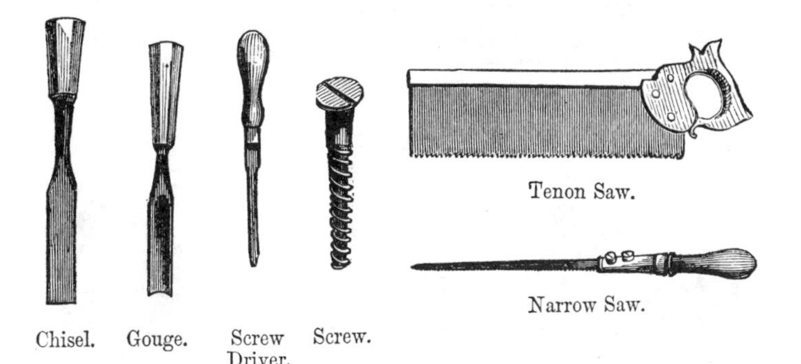

Chisel. Gouge. Screw Driver. Screw.

Tenon Saw.

Narrow Saw.

The *tenon saw* is used for cutting tenons, or flat slices from the ends of beams, that other beams cut in the same way may rest upon them and yet leave a flat surface. The thick back of the tenon saw keeps the blade from "buckling," or twisting; as it would be very likely to do while sawing in a horizontal direction.

The *dove-tail saw* is similar to the tenon saw, but smaller, and with a brass back instead of an iron one. It is used for dove-tailing, or cutting notches in a board or beam, into which projections in another board or beam are fitted, in order that the two may be held together, as we see the sides of a box are fitted to the back and front. Then there is the *compass saw*, for circular work, and the *keyhole*, or *narrow saw*, for cutting out holes.

The planes are used for bringing the edges and sides of beams, boards, or other wooden fitting, to a perfectly smooth surface: the first of these, used upon the rough wood, is

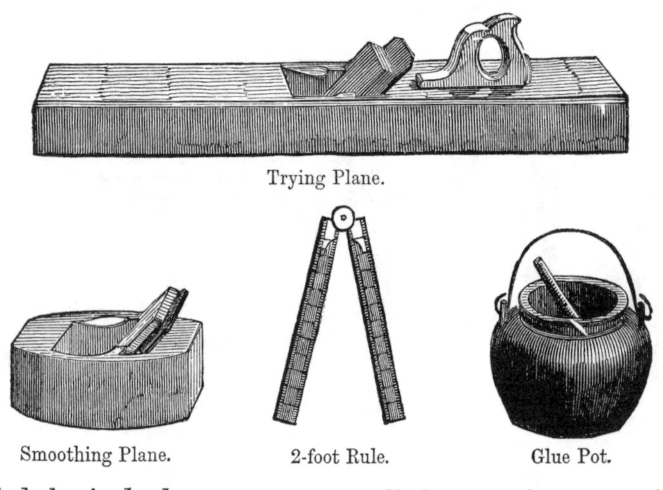

Trying Plane.

Smoothing Plane. 2-foot Rule. Glue Pot.

called the *jack plane*; another is called the *trying* or *trueing plane*; and a third, the *smoothing plane*.

The plane, as you will see, is a solid piece of hard smooth wood, with a hole in the centre containing the cutting tool, or, as it is called, the plane iron, which is firmly fixed with a wedge, so that its sharp cutting edge only slightly projects at the bottom.

The wood which is to be smoothed is fixed on the joiner's

bench by means of a screwed board, called a shooting board, and, by means of the handle at the top, the plane is made to slide swiftly along its surface, so that the edge of the tool cuts off a thin shaving.

There are various sorts of planes besides these, used for cutting various parts of the work, but they are most of them of similar construction.

Chisels are either for paring the wood, and are used with the hand; or are intended to cut into the thickness of the

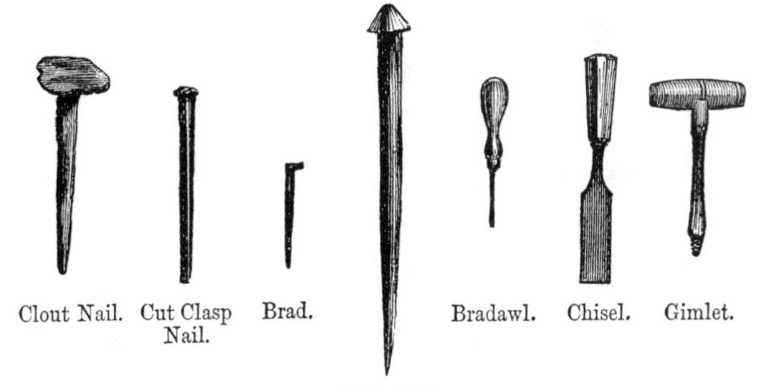

Clout Nail. Cut Clasp Nail. Brad. Bradawl. Chisel. Gimlet.

Spike Nail.

wood, and are then struck with a large wooden hammer, called the *mallet*. The *gouge* is a curved chisel, used for cutting mouldings or making round edges.

The boring tools are the bradawl, the gimlet, and the stock and bit.

The *bradawl* is a small sharp wire fixed in a handle, and used for making holes to receive large nails, which, if driven in at once by the *hammer*, would split the wood. Nails are of various kinds, the difference in which may be seen in the engraving. The best kinds of nails are made from thin bar iron, pointed, cut off to the proper length, and the head formed by stamping.

The *gimlet* is a hollow blade with a screw at the end, and fixed to a cross handle. It is used for boring a larger and rougher hole than the bradawl, for receiving *screws*, which are screwed into their place by the *screw driver*, a sort of chisel, the edge or point of which enters the notch in the head of the screw, so that the workman may turn it round.

The *bit* is a tool not unlike a large bradawl, which fits

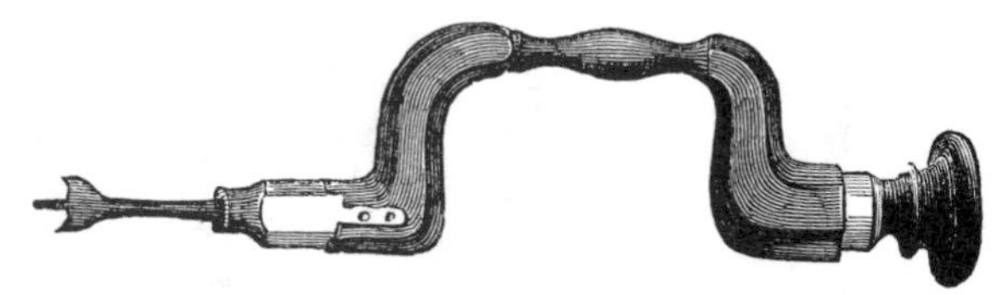

Centre Bit.

into a stock or handle. The bits are of various sizes, and are used for boring large holes.

The *brace and bit*, or *centre bit*, is a tool with a centre and two sharp points, one on each side; it is placed in a bow-shaped stock with a round loose end. This loose end the workman holds firmly, while he places the point of the tool against the wood required to be bored; he then turns the bow briskly round, and the two points revolving rapidly, cut out a circular piece.

The *pincers* are used for removing nails from wood, and it is easy to see how they are applied. There are some other tools which belong to the trade of the Joiner, but they are used less frequently, and a description of them will be given when we have to say something of the trade of the

Cabinet Maker, whose work resembles the finer part of that which the Joiner does at his own workshop.

The bench (*see large cut*) is the great table of thick planks supported on a timber frame, at which the Joiner

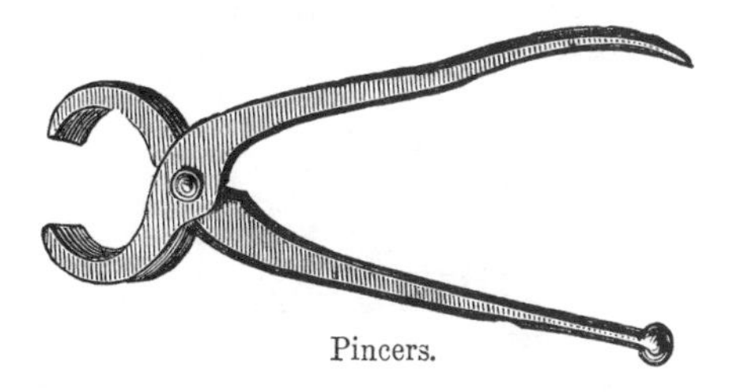

Pincers.

works; it is furnished with a "side-board" perforated with holes to receive a pin, against which one end of the wood on which he is operating rests, the other being firmly fixed in the bench screw.

THE HOUSE PAINTER AND GLAZIER.

HOUSE-PAINTING AND GLAZING.

ALTHOUGH we have spoken of the trade of the Plumber separately, and have placed the Painter and Glazier together, it generally happens that the three trades are carried on by the same persons, and you will often have noticed "Plumber, Painter, and Glazier" over the fronts of shops, where leaden pipes and taps, casements, and squares of coloured glass, and specimens of "graining," or imitations of various woods in painting, are placed for show in the windows, to represent the different businesses carried on.

It is the duty of the House Painter to cover with his colours such portions of the Joiner's, Smith's, or Plasterer's work as require to be protected from the action of the air, which would cause them to rust or decay. He has also to choose the colours which will be best suited for decorating walls and cornices, and generally to apply the proper shades for all the ornaments of the house, and this part of the business, which is called "decorative painting," requires a great degree of skill and taste in the workman, who may properly be called an artist.

The materials used by the Painter are principally white lead, linseed oil, spirits of turpentine, "dryers," and putty. White lead, which forms the basis of almost all the colours used in house painting, is prepared by exposing strips of lead to the action of acid; but the Painter buys it ready made, in the form of small cakes, or lumps. This white lead is used in all the under coats of paint, and generally makes the body of most of the colours which are afterwards applied. Unfortunately, while it is in the half-fluid state, when it is used by Painters, it is very poisonous and unwholesome, and many workmen suffer severely in their health from its use; but it will be found that this is often caused by their own carelessness in working, and by the want of personal cleanliness. The Painter should not only thoroughly wash his hands before every meal, but as soon as his work is done should entirely change his clothes, and wash hands and face thoroughly, or even take a bath. Beside this, he should wear, over his working clothes, a coarse linen frock or blouse, which will protect him from the spots of paint, and may be frequently washed.

Linseed oil and spirits of turpentine are used for mixing and thinning the colours; linseed oil is obtained from the seeds of the flax plant, which are heated, beaten, and pressed by machinery, until the greater part of the oil is extracted; the seeds crushed into cakes are then used as food for cattle. Nut oil, or the oil pressed from various kinds of nuts, such as walnuts, hazel and beech nuts, is also frequently used for mixing colours where they are likely to be exposed to the weather, but linseed oil is cheaper and is most generally adapted to the purpose.

Oil of turpentine, which the Painter calls "turps," is the oil obtained by distilling crude turpentine; and is used by the House Painter to make his colours work more smoothly and freely in the brush, to cause them to dry more quickly, and to take away that shiny unpleasant glare that would otherwise be seen on the surface of paint mixed with linseed oil alone.

It is necessary, too, to take some means to make the linseed oil dry more rapidly than it would in its raw state; or the paint would remain moist and greasy upon the walls for a very long time. For this purpose the oil is boiled, and is then known as "boiled oil," this is bought ready prepared, the Painter having by him preparations of lead or vitriol, called "dryers," which he mixes with his colours, after they are made, in order to increase the rapidity with which they set upon the places where they are applied.

The Painter has only a few tools, and these are very simple; they consist principally of the grinding stone, or *slab* and *muller*, for grinding his colours; *earthen pots* for holding the paints; *cans* for oil and turps; a few tin pots, or open cans and kettles, for colour to be applied to outside places or walls where he has to work on ladders; a *palette*

Slab and Muller. Paint Pot.

Palette Knife. Oil Can. Paint Pot. Paint Kettle.

knife, for spreading the white lead on the stone, or removing the paint to the palette, for fine work, or to the paint pot; and *brushes* of different sizes.

The various colours, which are mostly made from earths and minerals, are purchased by the Painter in powder or small dry lumps, and have to be brought to a fine state, and made quite free from grit, before they can be used.

The slab, or grindstone, is generally of marble or porphyry, and must have a perfectly smooth surface; and the muller is a large oval or egg-shaped pebble, with one end broken off, and the surface made as smooth as that of the slab. A small quantity of the colour which requires grinding

Paint Brushes.

is placed on the stone, and moistened with a little oil; and the muller is then worked over it, by a circular movement, until it is gradually driven to the edge of the stone. The colour is then removed by the spatula, or palette knife, and placed in the paint pot.

In this state, however, the colour is too thick to use, and of course requires to be thinned by adding the oil until it is sufficiently fluid.

In painting woodwork, the first business is called "knotting," that is, removing the turpentine from the knots, which would otherwise ooze and spoil the paint. They are first covered with fresh slaked lime, which dries up and burns out the turpentine; this is afterwards scraped off, and the knots painted with a mixture of red and white lead mixed with glue size, and afterwards with white lead and linseed oil. When dry, they must be rubbed smooth with a piece of pumice stone. The next operation is to put on the "priming," which is the first coat of paint, composed of red and white lead and linseed oil. The nail-holes, and other imperfections, are then stopped up with putty (a mixture of linseed oil and whiting, made into a paste), applied with the *stopping knife*; and then the other coats of paint are laid on; three coats being generally considered sufficient; and the last being of the required colour.

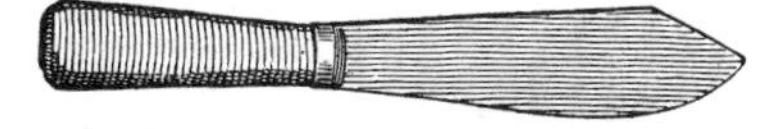

Stopping Knife.

The brushes used by the Painter, are either round or flat, the latter being used in varnishing or graining; they are made of hog's bristles. The smaller kinds of brushes are called *tools* or *fitches*, and are used for small surfaces, such as ornaments and mouldings round panels, when the colour is generally taken from a *palette*—a round slab of wood, with a hole through which the thumb of the left hand passes in order to hold it. The brushes must never be allowed to get dry, but when not in use must be kept in water: it is always best to keep a brush for each colour. Another part of the Painter's business which requires more skill, is graining and marbling; or the imitation in colours of the marks and grains of various kinds of wood, and the veins and spots in marbles. This is of course a part of the trade which requires a knowledge of the thing to be imitated, and great care and taste in doing it well.

Graining is generally done with "distemper colour," that is a kind of paint in which whiting and size are used instead of white lead and oil, for the basis of the colour; but the colours depend on the sort of grain to be imitated, and they are applied in various ways and with different sorts of brushes, some of the colour being occasionally removed with a piece of wash leather, in order to give the appearance of the light marks seen in the "heart" of the wood. The appearance of the "grain" of the wood is effected by the *graining comb*, a comb with short thick straight teeth, which is drawn along the paint in a wavy line.

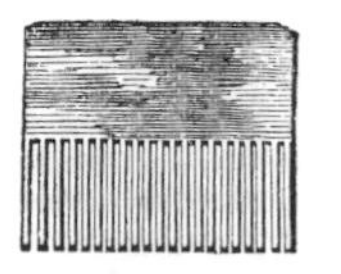

Graining Comb.

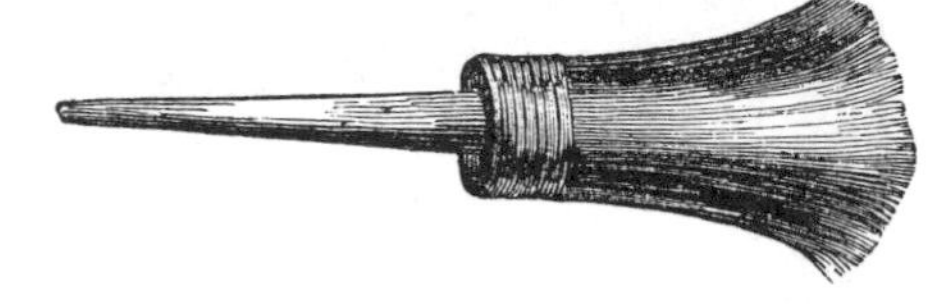

Painter's Brush.

Marbles are imitated in oil if it is outside work and exposed to the weather, but for inside work distemper colour is frequently used.

Occasionally walls are coloured in distemper, and it then frequently happens that a pattern or some ornamental design is painted in the centres of the panels, or round the mouldings. This is a part of the Painter's trade which requires much skill, and a knowledge of artistic drawing and design.

Varnishing is an operation requiring great care to perform it properly, since it is necessary not only to choose the right kind of varnish, but also to apply it to the surface of the paint with a light but firm hand. Varnishes are made of gums or resins melted, spirits of wine, oil of turpentine, or strong white drying oil. The hard varnishes dry rapidly, and are made of the harder kinds of gums, such as copal, mastic, &c., and the soft of Canada balsam, elemi, turpentine, &c. The most useful for the House Painter are those of copal, linseed oil, and turpentine.

The brushes used in varnishing are generally flat, so as to enable the workman to lay an even surface on the work. Varnishes are usually kept in wide-mouthed bottles; from which they are poured into little tin pans with a false bottom above the real bottom, the space between the two being filled with sand. The use of this is that when the pan is placed over the fire, the sand becomes heated, and the varnish is kept a long time from becoming chilled.

THE GLAZIER.

The trade of the Glazier, though now very important, was unknown in this country till the eleventh century, and even long afterwards the use of glass for windows was extremely rare; pieces of horn, and oiled paper, supplying its place in almost every building except palaces and churches. As we have here to do with the Glazier, and not with the manufacturer of glass, we need say little about the method of making the materials used; so that it will suffice to know that the "crown glass," mostly sold for windows of houses, is composed principally of white sand, pearlash, and saltpetre. This glass the Glazier purchases in sheets of a circular form, each of which is called a *table*, since the mode in which it is made renders such a shape

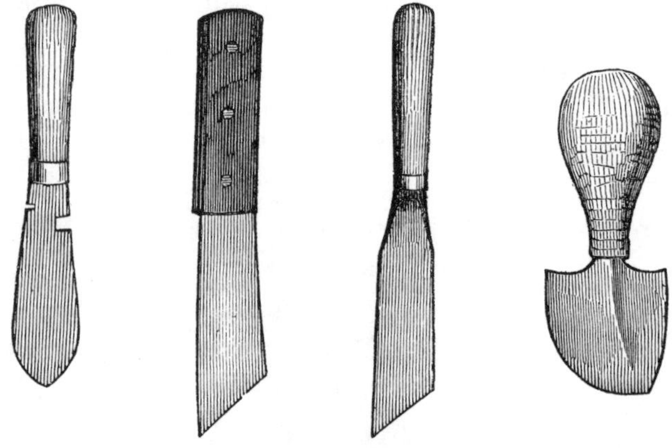

Glazing Knife. Hacking Knife. Chisel Knife. Clicker's Knife.

necessary; and the first thing he has to learn, is how to cut out square pieces of the proper size without wasting much of the round edge. The glass when cut is fixed either into lead work or sashes; the former of these is the oldest description of glazing; and in the common kind the leaden frames are soldered together, so as to form squares or diamonds; the sides of the grooves in which the glass is placed being soft enough to bend back to receive the panes, and then bend back again to hold them firmly. In wooden sashes, such as are now commonly used, the space or frame for each pane of glass is "rebated," that is, it has outside a small groove all round to receive the pane, and a ledge against which it rests. In this "rebate" each square of glass is placed, and then firmly bedded and fastened smoothly with a rim of putty filling up the groove.

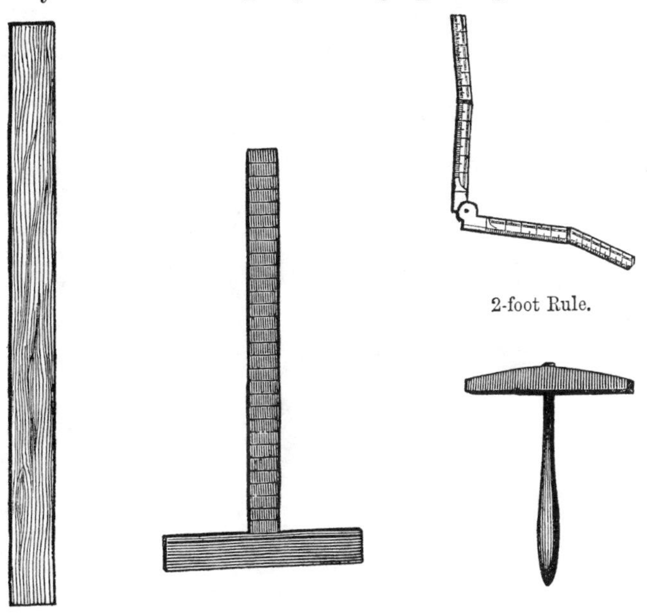

2-foot Rule.

Rule. Square. Tilter.

The Glazier uses a *hacking knife* for cutting out old putty from broken squares; and the *stopping knife*, or *glazing knife*, for laying and smoothing the putty.

For setting glass into lead work, the *setting knife*, or *chisel knife*, and the broad-bladed knife is used.

As much of the Glazier's work is done outside the house, he is sometimes compelled to use a board which is fastened with screws, and projects from the window so that he may sit astride on the outside. No one should ever attempt this where the sash can be easily taken out and the work done inside.

The rest of the Glazier's tools are a *square,* and a *straight edge rule* for cutting against, a *two-foot rule,* and compasses for measuring; *pincers* for breaking off the edges of glass that have been partially cut through; and the *diamond,*

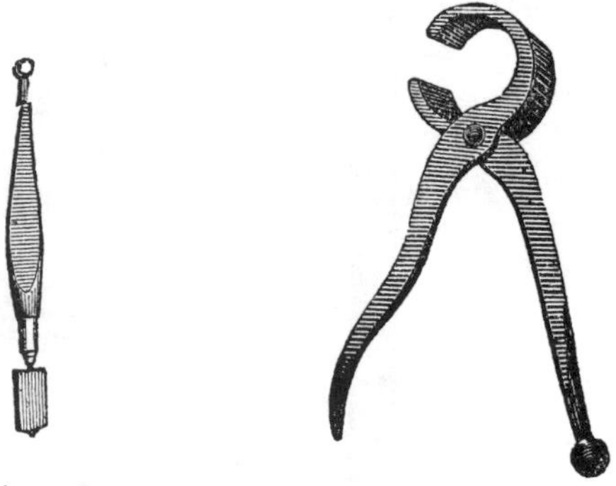

Diamond. Pincers.

which is used for cutting, and is his principal tool. This is made of a small unpolished diamond fixed in lead, and fastened to a handle of hard wood.

THE PLUMBER.

Casting Lead.

In trades connected with building, the work of the Plumber is of so much importance that it must come next to that of the Bricklayer. and the Carpenter. At one time the Plumber (who takes his name from the Latin word for *lead**) was principally employed in making leaden roofs of churches

* Plumbum.

or large public buildings and in forming casements for windows; window frames being then made of strips of lead soldered or riveted together and holding the little diamond-shaped panes of glass between their edges. In those days all the water used in the house was carried from the well or from the conduit in the main street, or was brought in casks set upon wheels from the nearest running stream: while rain-water for washing was collected in tubs or vats as it ran off the roofs. Not much more than a century ago the poets wrote of the misery of the streets of London on a wet night, when there were no waste pipes to carry off the rain from the overcharged gutters on the tiles, and nobody could venture out of doors without being half drowned by the sudden discharge of a shower bath from some overhanging gable. Then, as there was no proper system of pipes for carrying off the wet, there was very little drainage except by means of open gutters, and the bye-ways, as well as some of the principal thoroughfares and large houses, were extremely unhealthy.

We are not quite perfect even yet in these respects, and there are still neighbourhoods in London where a few Plumbers might be able to make vast improvements; but we are a great deal better off than our great grandfathers were. The Plumbers do not make quite so many leaden casements as they made in the olden time, but they are well employed in constructing roofs; carrying water into houses by means of leaden pipes; making cisterns to contain a good supply of it, and providing other pipes and gutters for conveying all the dirty water and drainage into the sewers that are under the roadway, where it runs quite away from the streets, and (in London at least) goes into the sea from near the mouth of the Thames at Erith.

Now as the Plumber has to do a great deal of his work in roofs and other places where he is liable to fall, he should be clear-headed, and able to go up a ladder or look over a ledge at a height from the ground without being afraid, and this is all the more necessary because he has to handle tools, and sometimes to pour melted solder out of a hot *iron ladle* while he is at his work on these places. For the purpose of melting the solder he has to carry his *fire grate and melting*

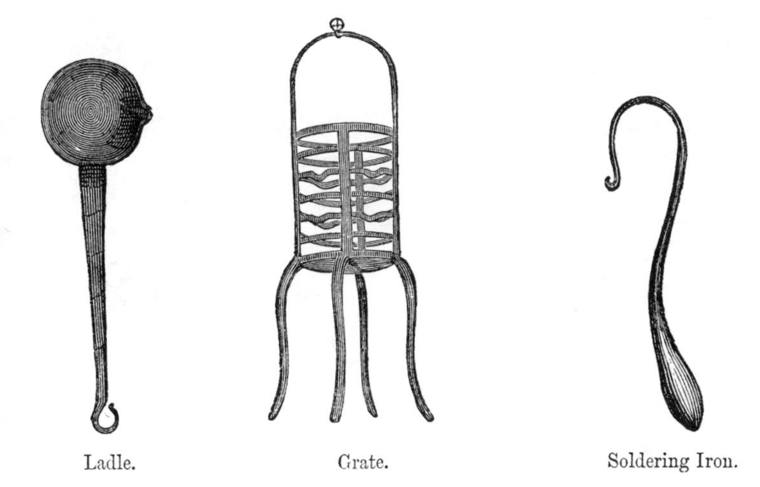

Ladle. Grate. Soldering Iron.

pot with him to some place near where he is at work, since unless the solder can be used rapidly it cools, and will not make a sound joint. The rest of the Plumber's tools are the *pouring stick* for applying the melted solder in the places where it is required, the *soldering iron* which is made red-hot, and passed over the soldered joints to smooth them and make them all firm and sound; the *chisel, shave hooks, drawing knife,* and *chipping knife,* for cutting the lead and scraping it on the surface, or at the edges that are to be fastened together; the *hammers* and *mallets* for beating the lead into shape and flattening the ridges, the *bossing mallet* and dresser for bringing the sheet of lead to a proper shape, and forming it over the ridge of a roof, the *chased wedge*, the screw-driver, the *dunring* and the *turnpin* for various uses in making roofs, laying down leaden pipes and fixing taps,

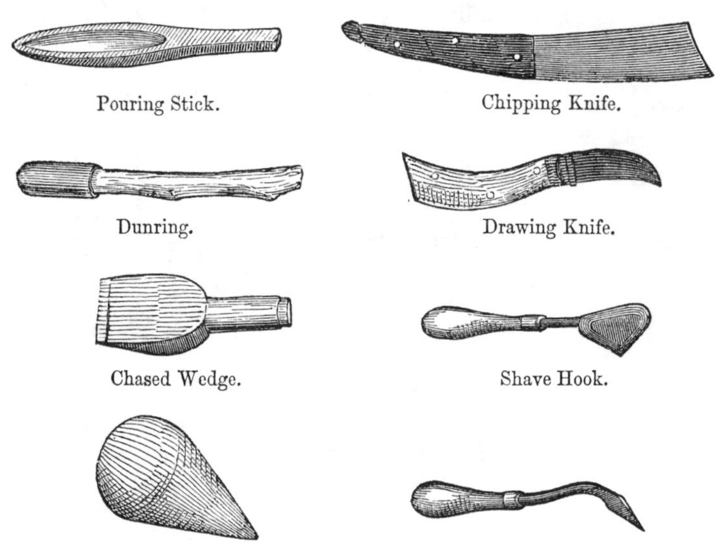

Pouring Stick. Chipping Knife.

Dunring. Drawing Knife.

Chased Wedge. Shave Hook.

Turnpin. Shave Hook.

and the *sucker hook* used in repairing or fixing pumps when the part of the pump called the sucker requires to be rectified.

There are other tools beside these, such as planes for making the surface of the lead smooth and even, gouges and centre-bits for circular openings in the lead to receive nails or clamps, measuring rules and compasses, and pads of carpet or cloth to hold under a pipe when it is being soldered, that the solder may be pressed round the joint before it cools, and without its dropping on the ground.

Plumbers now buy their sheet lead as well as their leaden pipe at the warehouses, but those in a large way of business formerly cast the lead themselves. For this purpose they used a casting table, which is a great wooden bench about six yards long and two yards wide, made of smooth

Mallet. Hammer.

planks, and with a raised wooden frame round the edge. On this table the Plumber spreads a layer of finely-sifted sand, which was made level by a strike—a flat piece of wood with two handles—drawn from end to end of the table; after this the surface was made still more smooth by a planer, which was a flat plate of copper fastened to a handle.

A trough called the pan ran along the whole length of the table, and into this the melted lead was poured from the melting pot. There were two ways of making the sheet lead: one of them was to tilt the trough up, and pour out the lead on to the table, two men immediately passing a wooden strike over it so as to spread it evenly over the whole surface. In this case the thickness of the sheet of lead depended on the distance between the edge of the "strike" and the surface of the sand. Another way was to

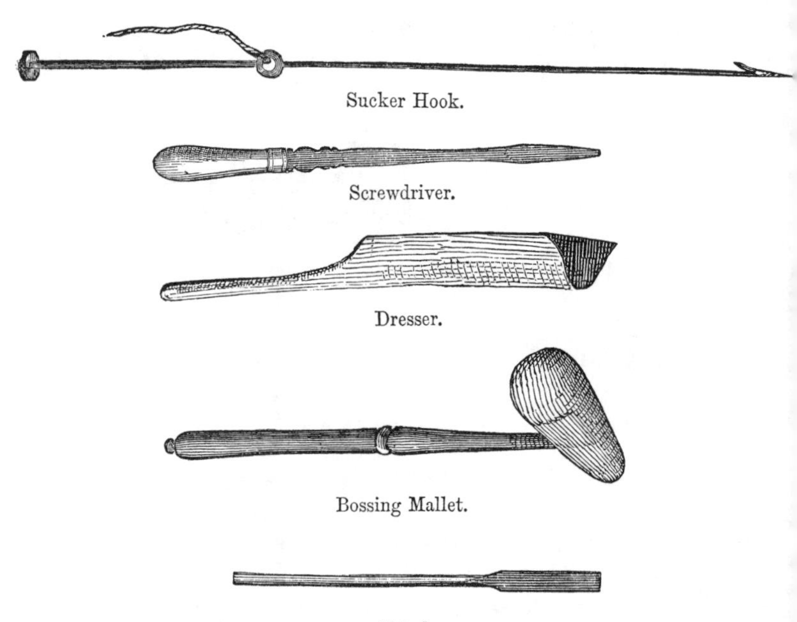

Sucker Hook.

Screwdriver.

Dresser.

Bossing Mallet.

Chisel.

have a narrow opening all along the bottom of the trough, the trough itself moving from end to end of the table as the lead flowed out. When this method was used the thickness of the sheet depended on the size of the opening, and the rapidity with which the trough was moved along the table.

Neither of these methods are now used, the lead being rolled into sheets by machinery.

Water-pipes are made by lead being cast in moulds with a steel rod passing through their middles, according to the size required. The lead is poured into the space between the rod and the mould. After it is cool the rod is drawn out by machinery, and the mould which is made in halves is opened and the pipe taken out. The pipe is then much thicker and shorter than is required for use, but it is afterwards drawn between powerful iron rollers with grooves cut in their surfaces, an iron rod being again placed inside it. As these grooves gradually decrease in size, and the pipe is drawn through several, it is very considerably lengthened, and at the same time diminished in thickness, by the time the operation is finished. Another method of making lead

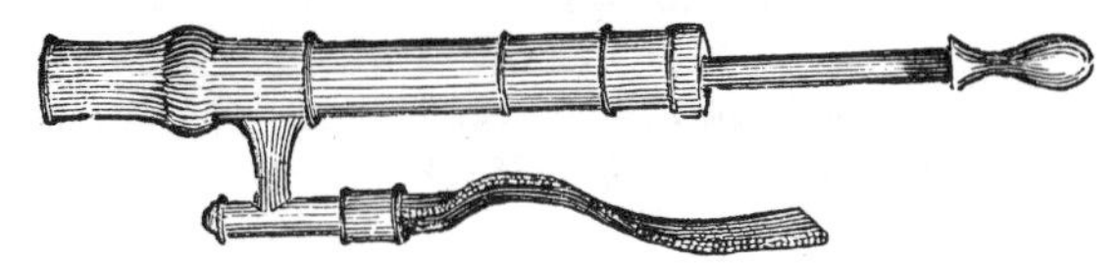

Forcing Pump.

pipes is by the use of the *forcing pump*, which pumps the melted metal out of the boiler into a mould containing a "mandril," or pipe of the required size. Pipes made in this way do not require to be rolled.

The work which the Plumber is called upon to do on the roofs of houses requires experience before it can be properly performed. The foundation of the roof which is to be covered with lead is made either of boards or plaster, so that the surface may be even, and if it be of boards they must be thick and well seasoned to prevent their warping.

The foundation slopes a little in order to carry off the rain towards one end. When the roof is so large that it needs two widths of lead, there are three ways of joining the edges of the lead together: one is by fastening to the roof long slips of wood (flat at bottom and round at top) at the places where the lead will be joined. Over these strips the edge of the first sheet of lead is folded and hammered down quite close, then the edge of the second sheet is folded and hammered over that, so that water cannot get between them. This is called "rolling."

Another method is to bring the two edges up just as though they were to be sewn together, then to fold them tightly one over the other, and hammer them down: this is called "overlapping," but it is not so good as rolling for keeping out the rain.

The third way is to *solder* the edges together: the solder, which is in constant use by the Plumber, being a metal made by mixing lead and tin together. These two metals when mixed adhere very strongly to the lead that they are meant to join, and the surface to which they are applied is made hot enough to unite with the solder by means of the soldering iron, while very often a little resin, borax, or tallow is placed on the surface of the lead to cause it to combine more rapidly with the molten metal.

Besides the fixing of roofs the Plumber makes leaden cisterns, fixes rain-water gutters, and waste pipes, and arranges taps and drains; and in some of these operations he will use zinc instead of lead. Not the least important part of his business, however, is the construction and arrangement of pumps, and for this purpose it is necessary that he should study mechanics, and those branches of science which refer to the properties of water, and to the laws which govern the air and other fluids. He is always the best workman who goes to his business with a knowledge of the natural laws and scientific facts connected with it, and a very little study will save a world of blundering; during which the ignorant man will remain a labourer, because it takes him half a lifetime to learn his business.

THE GASFITTER.

GASFITTERS AT WORK.

As there are now few large houses which are not lighted by gas, the trade of the Gasfitter is one of considerable importance; and though the materials used are generally bought ready made from the Brassfounders, the glass-works, the Ironfounders, or the lead warehouse, considerable skill is required, as well as some taste, in properly adapting the chandeliers and burners, and skilfully adjusting the tubes and pipes so that they may easily be repaired, or any escape of gas quickly detected. The various tools used by the Gasfitter are simple enough, but careful practice is necessary for their proper use; and, as any flaw or imperfection in the work may lead to very dangerous consequences, all the operations should be thoroughly tested, and every joint and fitting in the various parts made sound and strong.

Before the gas is taken into a house, or as it is called in the trade "laid on" in the house, it is of course necessary to obtain the permission of the Company to whom the works where the gas is made belong, and to agree to pay

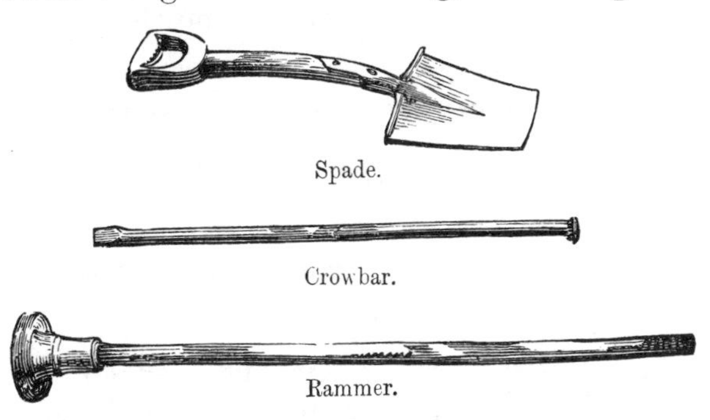

Spade.

Crowbar.

Rammer.

for the quantity that is burnt, which is charged by the thousand cubic feet, and varies in price according to the cost of the coal from which it is made, and the difficulty of conveying this coal to the works.

The first operation is to take up the roadway in front of the house, in order to connect the pipe which is to convey the gas to the meter with that which conducts it from the works to the different streets of the neighbourhood. For this purpose the *spade* and *crowbar* are necessary, while sometimes the pickaxe also has to be used; and the *rammer* serves to beat the earth down more closely after the pipe is laid.

The pipe which is joined to the larger pipe, or *main*, is generally of iron, and is made with a screw and socket, that it may not be easily displaced by the pressure of the roadway or footpath, and this, passing into the basement of the house, supplies the gas to the *meter*. The meter is a mechanical contrivance, which is so constructed that the quantity of gas passing through it is registered by a plate something like a clock face, with a hand to point to the figures which represent the number of cubic feet consumed. There are different kinds of meters, and the construction of

Brick Bit.

Brick Auger.

them varies very considerably, but they all answer this purpose: so that when the inspector visits the house once a quarter he may directly see what quantity is to be charged for. The pipe leading from the meter is fitted with a strong tap, by turning which, all the gas may be shut off from the tubes that convey it to the different burners in the house; and from this pipe the smaller pipes (made of a sort of solder or of lead) are taken to the various rooms. In order to carry the tubes through a wall, it is necessary to use the *brick bit* and the *brick auger* for boring a hole to receive it, while to support it against a wall or along the top of a ceiling the Gasfitter uses *wall hooks,* the *hammer* and the *tongs* for holding and bending the pipe to its proper direction. In kitchens and basement rooms the pipe is generally carried from the meter up the wall and along the ceiling to the place where the burner is to be fixed: but

Wall Hooks. Nuts.

Grease Pot.

Hammer.

in upper rooms the pipe is taken up the house wall, and carried under the flooring of the room above to the centre of the middle joist, where a hole is bored quite through the ceiling of the room below. For this purpose the *twisted*

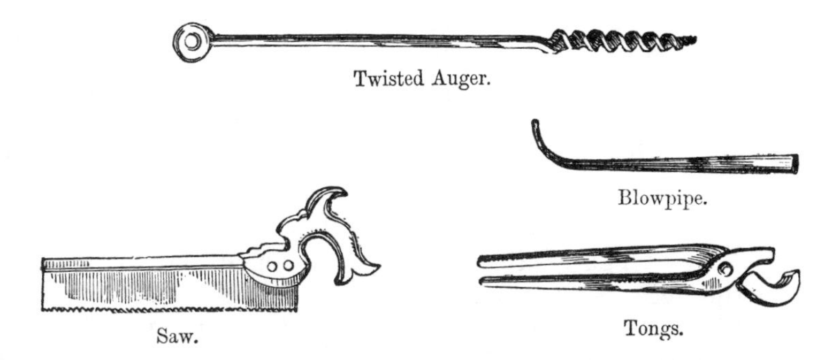

Twisted Auger.

Blowpipe.

Saw.

Tongs.

auger is generally used; and for the preliminary work of taking up the floor and cutting a groove in the joist to receive the pipe, that it may not be injured by the pressure of the boards above, the *saw* and the *hammer* are required.

Of course one length of pipe is not sufficient to go to any great distance, and a joint has frequently to be made, the solder used for this purpose being so easily melted and so readily combining with the pipe itself, which is of almost the same material, that the flame from prepared *tow* blown to an intense heat by the *blow pipe* is sufficient for the purpose. The rasp is used for filing the surface at the ends

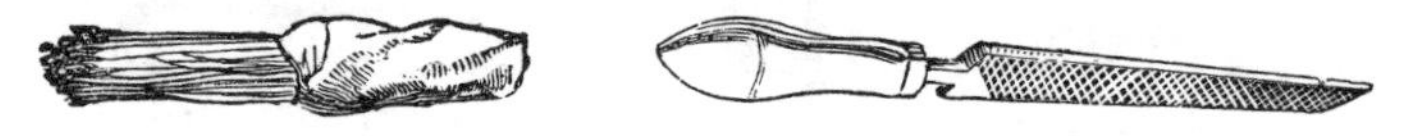

Waxed Rushes.

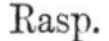

Rasp.

of the pipes where they are to be joined, and the *waxed rushes* to try the joints in order to see that there is no escape of gas through some small hole. When the pipes are all laid and brought to their proper positions the fittings are fixed. Those which descend from the centres of ceilings are generally called chandeliers, and consist of a tube of metal fastened to the gas pipe, the end of which coming through the ceiling is furnished with an iron screw to fit the end of the tube. This tube is placed within a larger tube, from the bottom of which a still smaller one passes within the first. The gas therefore descends the tube from the ceiling, and enters the smallest tube, which communicates with the *burners* that spring from the largest. The larger tube is intended to receive water, in which the end of the first tube rests, so that the gas is prevented from escaping, while by a nice adjustment of weights running over pulleys attached to the larger, the burners can be raised or lowered, as one tube slides within the other (*see large cut*). The *taps* are placed at the burners to turn on or turn off the gas as it is required, and in order to fix and screw these, as well as to fasten joints in the fittings by means of *screw nuts*, the *pliers* and the *wrench* are used.

The *clamps* are a sort of vice with grooved holes, for holding the taps and metal joints firmly while they are

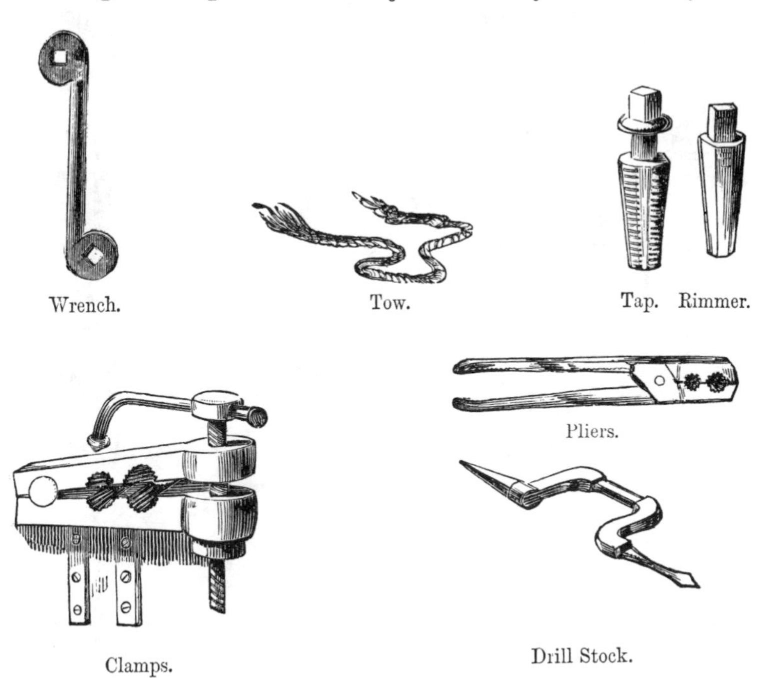

Wrench. Tow. Tap. Rimmer.

Pliers.

Clamps. Drill Stock.

filed or otherwise prepared; the *drill stock* is used for boring small pipes in order to make a branch to some other direction, and the *drill brace* is intended for boring the main pipes, under which the large hooks are placed while a stock and bit attached to the upper screw makes the hole; the ordinary *braces* are fitted to a *ratchet* or cog wheel at one end, and are also used for making incisions by being worked backwards and forwards.

The several kinds of *burners* are pierced or cut in such a way as to make the flame from the gas of different shapes, such as *fish tail, cock spur, star, or argand,* the names of

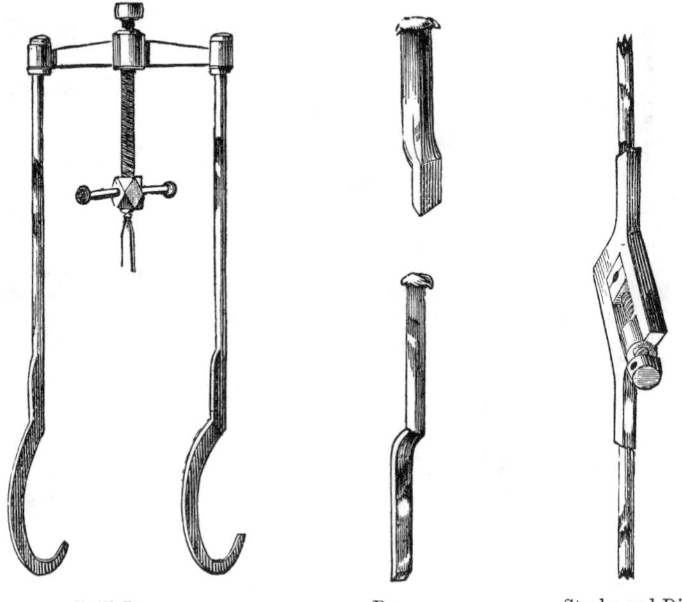

Drill Brace. Braces. Stocks and Dies.

which tell pretty well in what shape they appear. Small rooms are sometimes fitted with "telescopes" instead of chandeliers, the telescope being one tube sliding within another, and the space carefully filled with a properly adjusted cork or some other flexible and impervious subtance. Bedrooms, and apartments where light is required at the walls or chimney pieces, are furnished with brackets, which either simply project from the wall, or are made with an arm moving on a ball and socket joint. The manufacture of the various portions of the apparatus used in gas fitting will be mostly found described under the trade of the Brassfounder.

MANUFACTURE OF GAS.

GASOMETERS.

VERY few of the readers of this book will remember the appearance presented by London streets before the introduction of gas; when all the thoroughfares were darker than even the commonest streets are now, and the only light emanated either from the shop windows or from dim oil lamps, which the rain or the wind would frequently put out, even when they did not burn out of themselves for want of being trimmed and replenished. A century ago these oil lamps were quite insufficient to light even the main streets, and people who walked out at night generally hired a "link boy" to light them as they went, with a great flambeau of hemp and pitch, which smoked and smelt insufferably. The nobility and gentlefolks who rode in carriages were also attended by footmen with flambeaux of a better sort, and outside the doors of some of the oldest houses in London there may still be seen the great iron extinguishers attached to the railings, where the torch-bearers put out their lights till it was time to escort their masters and mistresses home. The poet Gay, who wrote the celebrated fables, describes the link boys, and gives them rather a bad character for so often being connected with the bands of thieves and footpads which infested London streets, and robbed people with impunity in the dark. He says :—

> "Though thou art tempted by the link man's call,
> Yet trust him not along the lonely wall;
> In the midway he'll quench the flaming brand,
> And share the booty with the pilfering band.
> Still keep the public streets, where oily rays,
> Shot from the crystal lamp, o'erspread thy ways."

Long before this, however, the inflammable nature of the vapour which streamed from burning coal had been observed by scientific men, and it was thought by many people that some method might be invented for making this gas useful for the purpose of lighting streets or houses. It had also been observed that the air suddenly escaping from the shafts of coal mines was often highly inflammable, and some experiments were made in the distillation of coal as early as 1726. In 1765 Lord Lonsdale proposed to the magistrates of Whitehaven to convey the gas from the neighbouring mines through pipes for lighting the town.

A number of eminent men afterwards made experiments with gas, but no decided practical result followed until a

Mr. Murdock, of Cornwall, began to manufacture gas for lighting his house and offices at Redruth. In 1798 the same gentleman used gas for lighting the Soho Foundry, where four years afterwards a public exhibition of the new invention was made by means of an illumination to celebrate the proclamation of peace.

In 1803 a gentleman named Winsor first publicly showed at the Lyceum Theatre, in London, a system of illumination by gas, which was the commencement of our present method of lighting our streets and houses; and after great difficulties and various experiments, a company was formed for the purpose of carrying out the undertaking of superseding the old oil lamps and making use of the new invention. The premises of the company and their factory were situated in Pall Mall, where the Carlton Club now stands, and the lights first appeared from the corner of St. James's Street to the Haymarket, while several jets were placed in front of Carlton House, the residence of the Prince Regent, afterwards George the Fourth.

By slow degrees, and by the assistance of Mr. Clegg, an eminent engineer, the manufacture of gas improved, and in 1814, when the allied sovereigns visited this country, and a general illumination was ordered, a magnificent pagoda in St. James's Park was erected for the exhibition of the new light. The following year Guildhall was fitted with gas burners; and, although it had been predicted that the new invention would *ruin the navy* by superseding the use of the oil brought by the whalers, and though some people declared that it was only a scheme for blowing up London, the success of these experiments led to its rapid adoption in most of the large towns of Europe.

It will now be necessary to say something of coal, of

which more than 400,000 tons are used every year for the manufacture of gas in London alone.

The name coal was originally given to any substance used for fuel, and the use of mineral coal, or, as it was formerly termed, *sea-coal*, from the fact of its being brought by sea, has not prevailed for longer than 200 years. Coal is found in this country in extensive deposits called coal-fields, the most important of which are:—1. Those of the great northern district, including the coal-fields north of the river Trent. 2. Those of the central district, including the Leicester, Warwick, Stafford, and Shropshire coal-fields. 3. Those of the western districts in North Wales, South Wales, and Gloucester. The coal occurs in a number of layers, or beds, termed *seams*, and these are separated from each other by layers of slatey clay, called *shale*, and coarse, hard sandstone, known as *grit*.

The seams of coal are mostly comparatively thin, but varying from a few inches to six or eight feet in thickness, and underneath the layers is usually found a bed of mountain limestone extending beyond the coal-field, and rising to the surface of the ground around it.

The first process for finding coal is to bore with iron tools a perpendicular hole in the ground; then if coal is discovered, a shaft, or pit, from ten to fifteen feet in diameter, is sunk, and lined with brick, cast iron, or wood, to prevent the admission of water and the falling in of the sides. When the shaft reaches a seam of coal, passages are made, twelve to fourteen feet wide, and varying in height with the coal; from these proceed smaller ones, which are again crossed by large ones, enormous blocks of coal being left to support the earth above. As the mine becomes larger, a second shaft is sunk at some distance from the first, in order that air may be supplied to the mine, and that the gas that escapes from the coal may be carried away. A current of air is caused by burning a large fire in one shaft, and the draught is directed to the part of the mine where it is required, by doors, which close up some of the passages and leave others open.

There are several varieties of coal found in this country. The most important are: the common, or bituminous coal, which soils the fingers on handling; the cannel, or candle coal, which burns with a bright flame, but does not soil; the anthracite, or culm coal, which burns without flame, does not form cinder when half consumed, and is most useful for furnaces, in consequence of its intense heat and the absence of smoke. Although often classed amongst the minerals, coal is evidently of vegetable origin; many sorts of it being distinctly fibrous, and showing the grain of the wood from which it is composed. Upon examination with the microscope, coal shows vegetable remains; and these are frequently so perfect that they prove its formation from such plants as ferns and fir-trees, of a kind somewhat different from those which now exist.

The value of coal depends entirely upon its inflammable and combustible properties. It is used as the source of artificial warmth in our dwellings, and our manufactures are mainly dependent on it. Without coal there would scarcely be any working in iron, copper, lead, or other metals, as in populous countries the supply of wood is soon exhausted. Our potteries and our glass works are also carried on by its use, and the power of steam which moves all our great engines and machines is dependent on coal for its existence.

Great Britain produces more than one-half the coal that is consumed in the world, and about 180,000 persons are employed in obtaining it in our collieries.

The first process in the manufacture of gas is to subject the coal to a great heat, by which it is *carbonised*, or burnt until only the cinder, or coke, remains after the gas has

Retort House.

flown off. This is done by placing the coal in retorts of fire-clay or iron, which are previously heated. These retorts, which occupy a building called the *retort house*, having been charged with coal, are perfectly closed, and the door *luted*, or stopped, with a sort of cement, so that the gas can only escape up the *ascension pipe*, the coke being left in the retort.

The gas from the coal then passes through the ascension pipe by what is called the *dip pipe*, into a main, a large horizontal tube extending along the length of the furnaces. This main is about half filled with water or tar, in which the ends of the dip pipes are immersed, so that as the gas runs in, it ascends through the liquid into the space above, but cannot flow back again into the dip pipes. In this way it is all collected in the upper part of the main pipe, and is ready for purification.

Where double retorts are used, each end is worked with at least three stokers, and an extra man for preparing the lids of the mouth-pieces. Others are required for extinguishing the coke, wheeling the coal into the retort house, clinkering furnaces, and attending to fires. Three stokers, assisted by a man to extinguish the coke, will perform all the work of taking off the lids, raking out the coke, extinguishing and wheeling it away from a bench of seven retorts, in twelve or thirteen minutes; they will then put the proper charge for each retort in the *scoop*, deliver its contents, and be ready for charging another bench in a further space of seven minutes, while a fourth workman will in the meantime have put on the lids, so that the whole work of discharging and charging the seven retorts will occupy barely twenty minutes.

This extreme dexterity is of course only acquired by long practice, and it must be admitted the labour is very severe; but this is moderated by the time the men have for repose between the charges. The first process in discharging or drawing is for one or two of the men to relieve the screws of the mouth-pieces of the retorts about to be discharged, by giving three or four rapid turns; another man instantly gives a knock to each of the cross bars to disengage them from the ears of the lid, and at the same time strikes the lid a blow with a piece of iron or hammer, in order to break the luting, and a light is immediately applied to prevent explosion, which would be likely to crack the retort if of clay. For want of this precaution, many lamentable accidents have happened through the gas exploding when combined with atmospheric air. The men then lift off the cross bar and screw of each retort, placing them on the ground, and then each seizes hold of a lid in both hands, lifting it by the projecting ears, and placing it aside to cool, ready for luting for another charge.

Fender.

Wheelbarrow.

Bus.

Three of the stokers then take up their *iron rakes,* which are simply rods of $\frac{3}{4}$-inch iron, about 12 feet long, having a handle at one end; the other end being turned at right angles is flat, about 6 inches long, 2 inches wide, and $\frac{1}{2}$-inch thick. These are inserted in the retort, and the red-hot coke drawn to the mouth, whence it drops into the coke vault, where there is a man ready to extinguish by throwing water on it; or when there is no vault the coke drops into *iron barrows* placed ready to receive it, and wheeled rapidly away when the charge is withdrawn. If the coke were not immediately extinguished it would smoulder, and the surface become covered with earthy ash, and detract from its appearance and value.

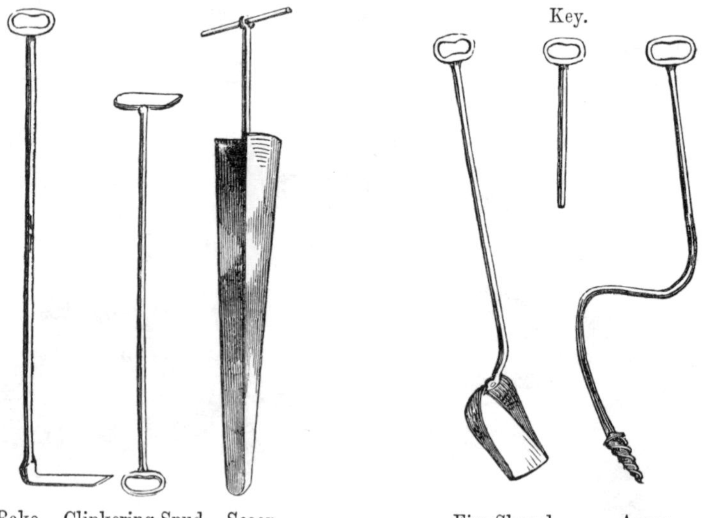

Rake. Clinkering Spud. Scoop. Fire Shovel. Auger.

Formerly, in charging retorts, the operation being comparatively very protracted, there was a considerable loss of gas, in addition to the time and extra fatigue to the men. In order to remedy these inconveniences, a method has been contrived for depositing the whole charge in the retort at once; for this purpose an *iron scoop* is used, this being a semi-cylinder of sheet iron, from 8 to 10 feet long and 10 or 12 inches diameter, with a cross handle at the end to assist in lifting and turning it round to empty the coals in the retort.

The charge of coal is placed in the scoop while it rests on the ground, having a bent rod underneath for the purpose of lifting it: one man takes hold of the cross handle, and two others lift the other end by the bent rod, and introduce it into the mouth of the retort. The scoop with its contents is then pushed forward to the further end, turned completely over, and immediately withdrawn, leaving the coal in the

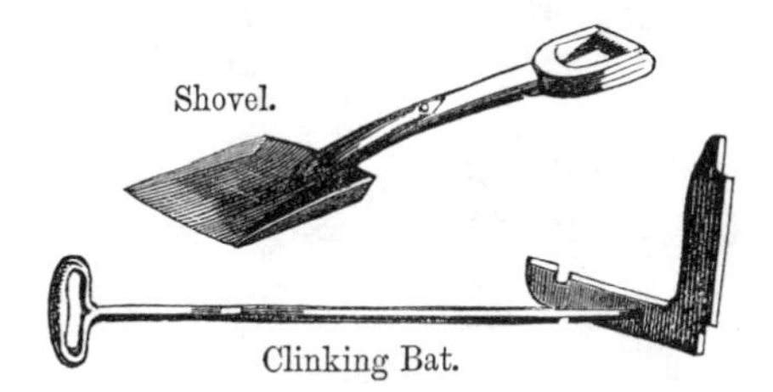

retort, which is raked into a layer of uniform thickness, when the lid, previously luted and ready, is placed in its position and screwed up as quickly as possible. The operation of charging a retort with the scoop does not occupy more than thirty or forty seconds, so that very little escape of gas can take place. The *shovel* is used for lifting the coal to the scoop, the *clinking bat* for breaking or removing the coke in the vault, the spud for a similar purpose.

The gas in the hydraulic main is of course very impure, having undergone no alteration since it came from the coal in the retort. It contains a quantity of tar and ammoniacal liquor in vapour, and these have to be separated from it;

and this can be effected by what is called condensation, the instrument used being known as a *condenser*. There are two or three forms of condensers in use, but a common one, which is represented in the engraving, is called the horizontal condenser, and is a rectangular box or chest formed of cast iron plates, put together with flanges, and perfectly tight joints. Its interior is provided with a series of iron trays, containing each about two inches in depth of water,

Condensers.

and so arranged that the gas, entering at the bottom of the chest, passes in succession over the surface of the water in each tray, and traversing the whole length of trough ten or twelve times, passes off at the upper side. In the mean time a continuous stream of water enters at the top, and in its descent absorbs a portion of the ammonia, at the same time cooling and condensing the vapours in combination with the gas.

Horizontal condensers are sometimes composed of a series of pipes placed in a horizontal position, and immersed in water.

The other form of condenser, which is very generally employed, consists of a series of vertical pipes, connected in pairs by semicircular bends at top, and attached to a cast-iron box or chest at bottom. This chest has a series of divisions, the ends of which are sealed by liquid placed

Wet Purifier.

therein, so that the gas in its passage has to pass through the whole series of pipes. The pipes by their contact with the atmosphere radiate the heat acquired from the gas in its passage, and it being in consequence cooled, deposits the vapours as liquid in the form of tar, and water saturated with ammonia, generally called ammoniacal liquor. This condenser is sometimes used with an application of cold water on its exterior, in order to increase the cooling effect.

The other part of the purification is by a chemical process, in which a solution of lime is used to remove other impurities in the gas. This is called *wet lime purification*, or when the lime is only slackened (or moistened) it is called *dry lime purification*, and the process takes the sulphuretted hydrogen, the carbonic acid, and other matters from the gas before it is stored for use.

Dry Purifier.

The wet lime purifier consists of a cast-iron cylinder entirely closed at top and bottom, except where the inlet and outlet pipes join it, and where an opening is required for charging it with lime-water, which same opening is also used for drawing off the charge. To the inside of the cover of this outer cylinder is bolted an inlet cylinder usually made of wrought-iron plate. This inlet cylinder is open at the lower part, and reaches to within a foot from the bottom of the outer cylinder, but has bolted to its lower flange a wide ring or dash plate of sheet iron, the outer diameter being only 8 or 9 inches less than that of the outer cylinder, so that a space of about 4 or 5 inches is left between the outside of the ring and the interior of the large cylinder.

The gas passes down through the inlet cylinder, and by its pressure forces its way up through the fluid lime, the surface of which is 8 or 9 inches above the dash plate.

The wet lime purifiers are variously worked; when four are used, two vessels are employed at one time, and when the lime in the first is incapable of absorbing the impurity, that purifier is put out of action, and the second and third are worked, and so on in succession.

When quick lime is slackened, reduced to powder, and slightly moistened with water, chemically this is called the hydrate of lime, and is often employed to absorb the sulphuretted hydrogen and carbonic acid from the gas. The process is termed dry lime purification.

Dry lime purifiers are generally rectangular cast-iron vessels, varying from 3 feet to 30 feet square, and from 3 feet to 4 feet 6 inches deep. Sometimes in small works they are made circular; this, however, is not very frequent, and is done for convenience or economy in construction. Each purifier contains a series of perforated shelves, trays, or sieves, supported by suitable bearers of wrought or cast iron, the ends of which are attached to "snuggs" cast on the purifier. In large apparatus there are also pillars placed at intermediate distances to carry the weight of the sieves and purifying material.

The upper part of the purifier is surrounded by a cistern or reservoir of from 6 inches to 24 inches deep, and from 3 inches to 6 inches wide, which is often cast with the purifier, and forms part of it, or at other times is attached thereto by bolts and cement, and is for the purpose of containing water to seal the cover. The cover of the purifier is of boiler plate or cast iron, the latter being preferable on account of its durability; but the increased weight is an impediment to its adoption. The rim or border of the cover is rather deeper than the cistern into which it is placed, and is effectually sealed by the water, so preventing the gas escaping from that point.

Often the purifier is divided into two compartments, so that the gas ascends through a set of sieves on the one side, and descends through another set on the other side, answering the purpose of two sets of apparatus. In all establishments, however small they may be, two distinct purifiers at least are necessary, to enable the impure lime to be removed from the one whilst the gas is being purified by the other.

The next subject for consideration is that of the *gas-holders*, or vessels in which the gas is stored ready for delivery into the mains, which distribute it throughout the districts to be lighted. These vessels were originally termed *gasometers*, which name is sometimes even now applied to them; but as they have nothing whatever to do with the measurement of gas, but are mere vessels of capacity or stores, the simple name of gasholder is more expressive and appropriate.

The gasholder is composed of two distinct parts, one of which contains water, and is called the tank, the other is the vessel which contains the gas, being really the gasholder. On the Continent the former is very generally termed the "cistern," and the latter the "bell."

The tank is a large cylindrical vessel, constructed usually, for the sake of economy, of brickwork or masonry, but when the ground is marshy, or when water exists abundantly a short distance below the surface of the earth, which would prevent the construction in masonry at a moderate price, these tanks are made in cast-iron, and, indeed, in small works, are often of wrought iron. In the interior of the tank there are two vertical pipes for the admission and egress of the gas, called the inlet and outlet pipes; the former being in direct communication with the manufacturing apparatus, the latter with the mains which convey the gas to the town. These pipes rise a few inches above the level of the top of the tank, so that the water cannot overflow into them. A series of columns, generally of cast-iron, but sometimes of wood, or brick piers, are placed at equal distances around the tank for the purpose of guiding the holder.

The holder is a cylindrical vessel closed at the top, which is termed the roof, and open at the bottom, made of sheet iron, varying in thickness according to the dimensions of the apparatus, the smaller sizes being constructed of thin material in order to avoid an excess of pressure, whilst those of very large dimensions are made of stout plates for the purpose of obtaining sufficient pressure to expel the gas to the burners. The holder is somewhat less in diameter, but of the same depth as the tank in which it is placed, sometimes being partially suspended by chains which pass over grooved pulleys and counter-balance weights, but more frequently only guided by rollers attached around its lower and upper edges, which work against suitable guides in the tank and on the columns in such a manner as to permit the holder to ascend and descend in the tank with the greatest freedom.

The action of the gasholder is very simple. The tank

being filled with water, and the holder immersed therein ready for use, there is a space between the surface of the water and the roof of the holder; the gas enters by the inlet pipe into this space, and with the force it acquires in being expelled from the coal, pressing on the surface of the water and underneath the roof, and over the whole area of both, causes the holder to rise. Thus, by its own force or pressure, the gas provides room for itself, and in proportion to the quantity entering so does the holder rise out of the water. For instance, a holder having 100 feet area, or about 11 feet 4 inches diameter, in rising 10 feet will receive 1,000 cubic feet of gas, and in descending, the same quantity would be expelled.

Gasholders, though often suspended, are never entirely counter-balanced, having always sufficient weight to give the necessary pressure for forcing the gas through the mains and smaller pipes to the burners, all through the neighbourhood which is supplied from it. The gasholder should be so constructed that, when it is full or at its greatest height, its lower edge will be so far under water as to prevent the gas from escaping.

The water in the tank serves three purposes; it is the means of resistance for the gas to lift the holder, it prevents the gas escaping or mixing with the atmosphere; and it is the means of expelling or forcing out the gas as the holder descends.

THE IRONFOUNDER.

FOUNDRY.

Having already described the various operations of the trades employed in building and fitting a house, we will say something of the manufacture of those cast iron columns, girders, gratings, balconies, pipes, gutters, air traps, coal plates, stoves, and other articles which are so necessary to the Builder before his work can be completed. All these, as well as a great variety of other goods made in black or bronze iron, such as gates, bridges, pieces of furniture (like umbrella stands), iron taps, and even pots and frying pans, are made at the *Iron Foundry.*

Iron is a metal of a bluish gray colour; but in its pure state it looks almost white when polished, and has a brilliant lustre, while when it is broken the broken portion looks dull and fibrous. It is the hardest of all the malleable and ductile metals, and the most tenacious of all metals, an iron wire of $\frac{1}{36}$th of an inch in diameter bearing a weight of 60 pounds.

In the pure state it requires the strongest heat of what is called a wind furnace to melt it.

Iron may be called the most precious of all metals; it is certainly the most beneficial to man, and its uses are innumerable; indeed, there is not a branch of human industry that could well afford to dispense with its aid and services; nearly all the tools, implements, instruments, and engines used by man are wholly or partly made of it, and we could better afford to give up all the other metals than to part with this, which is the most useful.

Iron is used in two different states, as *cast iron* and *wrought iron,* the differences between them depending on the proportion of carbon combined with the metal, cast iron containing the most and wrought iron the least.

For the production of wrought iron in the ordinary way, two distinct sets of processes are required; first the extraction of the metal from the "ore" that is brought up from the mine, which metal is cast iron; and secondly the conversion of this cast iron into malleable or bar iron, by remelting, *puddling*, and *forging*. Bar iron is turned into steel by placing it in contact with charcoal in a peculiar kind of furnace.

When the ore is taken from the mine it is first burnt or calcined, and then removed to a blast furnace to be smelted. These blast furnaces are generally built of brick, and look like small towers. The ore is mixed with limestone, which causes it to melt more easily, and the fire is lighted with pit coal or coke. The melted metal sinks to the bottom of the furnace in consequence of its weight, while the limestone and dross float on the top, and are allowed to run off when they cool into a mass of what is called "slag."

The melted metal is run off from the bottom of the furnace, either into moulds for some sort of castings, or into a large furrow made in a bed of sand. This large furrow has several smaller furrows on each side of it, and has received the name of the "sow;" the smaller furrows being called "pigs;"—and the iron when it is formed in this shape to be afterwards made malleable is called "pig iron."

The pig iron is taken to other and smaller furnaces called puddling furnaces, the bottoms of which are lined with clay mixed with the slag just mentioned, and forming a substance which the puddlers call "bull-dog," though it would be difficult to discover why it received that name.

About four hundred weight of the pigs is placed in the furnace, and as it melts the puddler stands at the furnace mouth with a long *iron rod* bent at the end, and stirs it about, until it comes to resemble several great balls of iron paste. These balls are removed, and fall into iron trucks pushed along a small railway by boys, who wheel them at once to the "shingling hammer," an immensely powerful hammer worked by steam, and this beats the iron into small square bars called "blooms."

The blooms are next carried to the rolling mill, which is a pair of great rollers cut into grooves of various sizes, and between these grooves the bars are squeezed, as the rollers turn round, until they become much longer and narrower, when they are known as "forged bars."

Some of the rolling mills, however, are plain cylinders without grooves, and when a slab of white-hot iron is placed between these it comes out from the pressure in a great broad sheet of metal.

These operations require great bodily strength as well as considerable skill on the part of the workmen, who are obliged to seize the heated metal with long tongs, and to catch it in the same way as it comes out from the mill.

The iron which is intended for castings is melted in a "cupola furnace," so called on account of the dome-like shape in which it is built, which has something to do with

Casting Ladle.

the more perfect heating of the metal. When the iron is completely melted so that it will run freely, the lower portion of the furnace is opened, and the white-hot stream is received in the *casting ladle*, or, where it has to be carried for some distance, and the casting is large, in great iron pails carried in a sort of frame by two men. From the casting ladle it is at once poured into the *mould*.

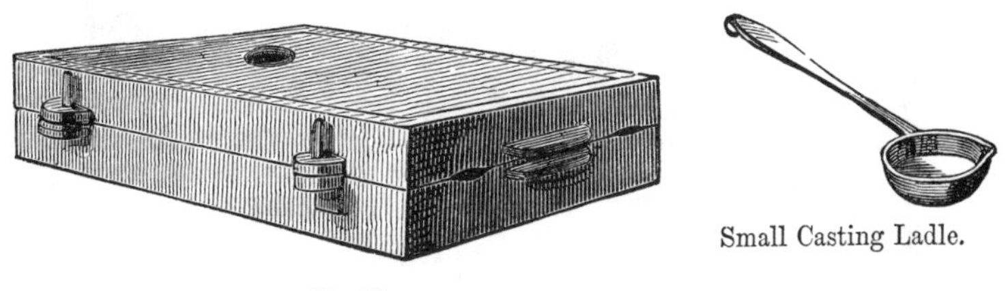

Small Casting Ladle.

Mould.

The mould is a sort of iron box filled with a peculiar sort of sand, into which a wooden pattern of the intended casting is pressed, and the sand firmly rammed down and made solid. There are, in fact, two boxes of sand, each of which is impressed with one half of the thickness of the required casting, so that when they are brought together, and firmly fastened with the pins, as shown in the picture, the patterns which have been taken out have left a half of the impression in each box, each corresponding exactly to the other. A hole in the box receives the melted metal, for which a channel has been left in the sand, that it may freely run into the hollows left by the pattern, and completely fill them; then, when it has sufficiently cooled, the casting is removed, and when the rough edges have been removed, and the irregularities trimmed off, it is ready for use, and may be fitted to its other parts, which have perhaps been separately cast, as in the case of garden seats, fenders, chandeliers, umbrella stands, or ornamental girders and columns.

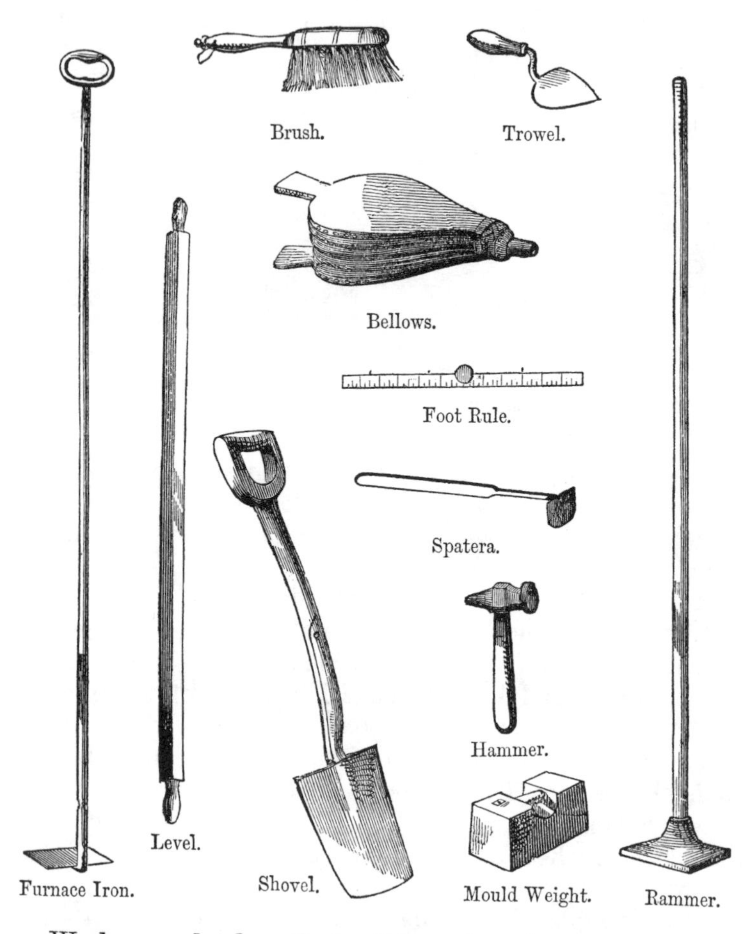

We have only described solid castings, but as ornamental iron work is generally made hollow, this has to be cast in rather a different way, though the only difference is that what are called "cores" are used. These cores are in fact solid metal patterns made a little smaller than the hollow left by the real pattern, and allowed to remain in the mould. The melted iron then flows between the surface of the core and the surface of the mould, and the casting is hollow, so that when the core is removed the metal is only the thickness of the space left between the two surfaces. You will see what is meant by placing a small teacup inside a larger one, and then pouring water between them.

The tools used by the Ironfounder are not very numerous: the casting ladles and mould have already been mentioned; the uses of the *shovel* and the *mould weight*, the *rammer*, and the *furnace iron* need no description.

The designers, and pattern makers, and the mould maker have the most important duties, and the latter will have to use a small *trowel* and a *spatera* for arranging his sand and loam, a *level* that it may be perfectly true, and a *brush* and a *pair of bellows* for removing any particles of grit from the surface of the channels where the pattern has been impressed.

Almost all irons are improved by admixture with others, and, therefore, when superior castings are required they should not be run direct from the smelting furnace, but the metal should be remelted in a cupola furnace, which gives the opportunity of suiting the quality of the iron to its intended use. Thus, for delicate ornamental work, a soft and very fluid iron will be required, whilst for girders and castings exposed to cross strain the metal will require to be harder and more tenacious. For bed-plates and castings which have merely to sustain a compressing force, the chief point to be attended to is the hardness of the metal.

Castings should be allowed to remain in the sand until

cool, as the quality of the metal is greatly injured by the rapid and irregular cooling whieh takes place from exposure to air if removed from the moulds in a red hot state, which is sometimes done in small foundries to economise room.

THE BLACKSMITH.

FORGE.

In the building and fitting of the house a large portion of the iron work will have to be furnished by the Smith, and as we have already given some description of iron founding, it will be necessary to say something about the Blacksmith, or the worker in iron or black metal, whose business is different from that of the whitesmith, which has to do more particularly with white or yellow metal.

The way in which malleable iron, that is, iron fit for the hammer, is produced has been mentioned in connexion with the trade of the iron founder, who in fact supplies the Blacksmith with the raw material. It is not very easy to tell you much about the way in which the Blacksmith makes the great variety of articles which his trade furnishes, for there is no business the success of which depends more upon personal skill. As the trade of the Smith, or at all events the worker in metal, is one of the most ancient, and existed in times when there were few tools,—as, in fact, it is the Smith who has to make tools,—so at the present day, he has to depend chiefly on his own ability in the use of the hammer and a few other simple instruments to fashion the articles that come out of his workshop.

It is he who supplies the various articles of wrought iron work used in a building; as pileshoes, straps, screw bolts, dog-irons, chimney-bars, gratings, and wrought-iron railings and balustrades for staircases. Wrought iron was formerly much used for many purposes for which cast iron is now generally employed; the improvements made in casting during the present century having caused a great alteration in this respect. It is not only for building purposes that the Blacksmith is employed, however, since there is scarcely anything constructed of iron in which his aid is not required, from important portions of machinery to the rough horse-shoes which have to be finished and fitted by the farrier. In the forge, where the great bellows suspended to the ceiling make the furnace roar, and the sparks fly, the clinking of hammers is heard all day long.

Hammer.

Set Hammers.

Tongs.

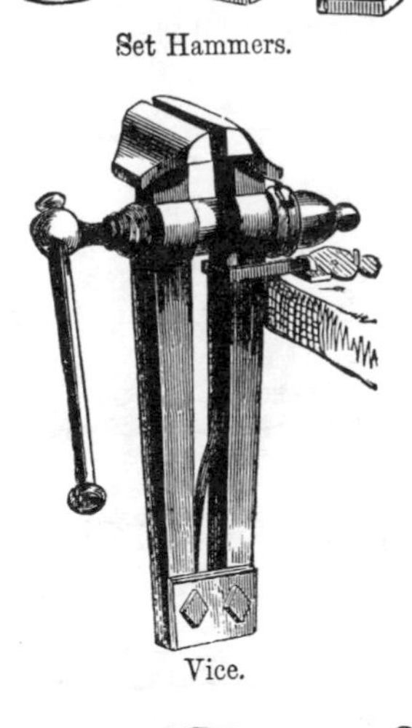

Vice.

Sledge Hammer.

Anvil.

File.

The *anvil*, on which the iron is beaten into shape, the bench, fitted with a *vice* for holding such portions of the work as require the *file*, the *tongs*, with which the red hot metal is held, the *sledge hammer*, and the *set hammers*, are the principal tools.

The sledge hammer is used for beating the metal until it is tempered and easily formed into shape, and it is in the

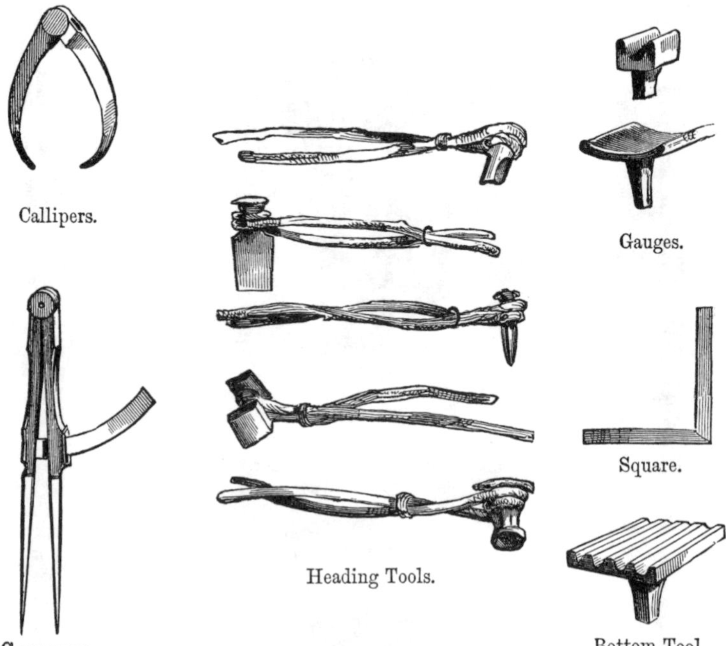

Callipers. Gauges. Compasses. Heading Tools. Square. Bottom Tool.

tempering of the metal by beating that the great skill of the Smith is often displayed. The set hammers are used for setting out the work, and have heads of different shapes, according to the form which the metal is required to assume. The various *gauges* are placed upon the anvil for the similar purpose of shaping the work, and the *callipers, compasses,* and *square,* measuring and adjusting it. The *heading tools* consist of cutting, punching, and stamping instruments, and are probably so called, because they are furnished with heads to receive the blows of the heavy hammer, by which they are forced into the hot metal on the anvil.

These heading tools are held, not by handles of their own, which would break off with the concussion of the hammer, but by a sort of withe of birch, or some other tree

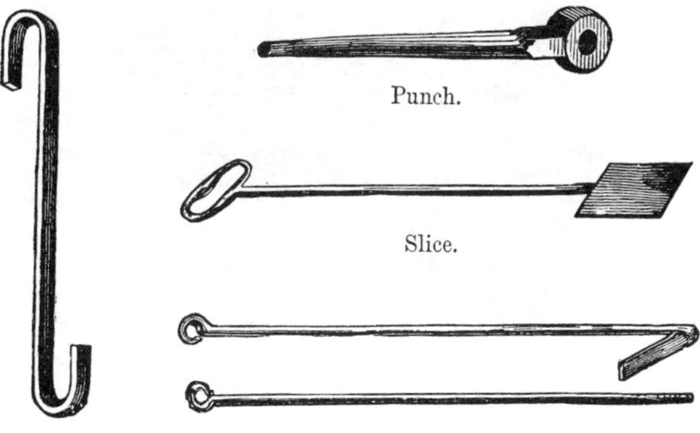

Punch.

Slice.

Double Hooks.

Fire Irons.

fastened loosely round them at their heads, and bound by a ring to keep them from parting. The *punch* is used in making bolts or rivets, the *slice* and *fire irons* for arranging the fuel in the furnace, and removing small articles after they are heated, the *double hooks* for removing or suspending bars, and for some other purposes.

THE BRASSFOUNDER.

Casting.

Next to iron, perhaps brass is the metal chiefly employed in the manufacture of articles of daily use, and the trade of the Brassfounder is therefore of very great importance, especially in connexion with the small metal fittings, such as catches, locks, bolts, hooks, screws, and other objects used in completing and furnishing the house.

Brass is not a pure metal, but is what is called an alloy, that is, a mixture of various metals. It is composed of copper and zinc in such proportions as may be necessary to obtain various degrees of hardness and colour, according to the use for which the compound is to be employed. The best proportions for common brass are about two parts of copper to one part of zinc. Formerly brass was made by heating copper with *calamine* (which is the ore of zinc) and charcoal, but it is now formed from melting the two metals together. It is then cast into plates, which are either broken up for recasting into any required form, or rolled into sheets. Common brass is very malleable, is more easily melted than copper, and may be cast into any form. It will take a very high polish, does not rust or tarnish by exposure to the air, and, although it is durable in wear, is sufficiently soft to

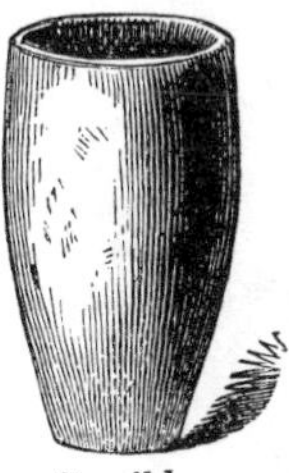

Crucible.

Magnet.

yield readily to the file and other tools used by the workmen. These properties make it useful for a great variety of purposes where steel or iron could not be so well employed.

The smelting or mixing houses where the brass is made are fitted with air furnaces, and in some of the best workshops the *crucibles* or melting pots are made of *plumbago* or pure blacklead, which, although it is more expensive, is much more durable than the Stourbridge clay, of which the commoner crucibles are formed.

A very fine quality of brass for best castings consists of

three parts of best selected copper, and two of spelter, with some best scrap brass and a little tin; while a second quality is formed of two parts of ordinary copper and one part of spelter, melted into ingots with a proportion of scrap brass and brass filings. Before the latter are used the iron filings are separated from them by a *magnet*, or by a series of magnets fastened to a revolving chain frame.

The sheet brass is procured from the mills where it is rolled, and the brass wire is also supplied from the special manufactories where it is drawn ready for use.

Making Moulds.

The method of *making moulds* for casting iron has already been mentioned, and those employed by the Brassfounder are quite similar, the tools used by the mould maker being the *trowel, mallet, rule*, and *sand hook*, the *shovel* for removing the sand, the *brush* for sweeping the surface, the *bellows* for blowing off the dust, and the *compasses* for measuring accurately.

The principal materials for making foundry moulds for brass castings are fine sand and loam mixed in various proportions, according to the nature of the work. New sand is

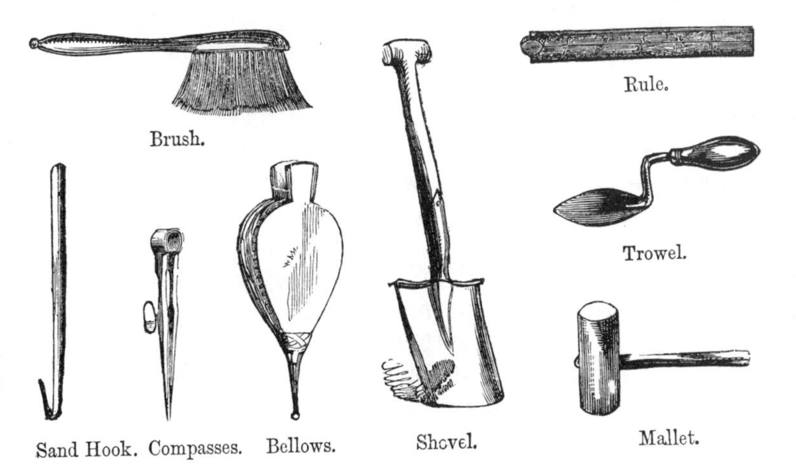

Brush. Rule. Trowel. Sand Hook. Compasses. Bellows. Shovel. Mallet.

used for fine castings, old sand for ordinary work. The requisite external support is given by a couple of shallow rectangular iron frames without tops or bottoms, called *flasks* or *casting boxes*.

The two halves constituting a casting box carry ears corresponding exactly with one another, one set pierced with holes, the other furnished with points entering truly into these holes, and which may be made fast in them by cross-pins or wedges. One of the flasks is laid face downward on a board longer and wider than it, and is then rammed full of moulding sand; the surface is struck off level with a straight metal bar or scraper, a little loose sand is sprinkled upon it, and another board of proper size placed over it and rubbed down close. The two boards and the flask contained between them are turned over and the top board is taken off; the clean surface of moist sand now exposed is dusted over with perfectly dry fine parting sand, or very fine red brickdust. The patterns or models are now properly arranged on the surface of the same, the cylindrical or thick parts being partly sunk in the latter, and care being also taken to leave sufficient space between the several patterns to prevent one part breaking into the other, and also passages or *ingates*, by which to pour in the metal and allow the air to escape. The patterns are arranged on both sides a central passage or runner, technically called a *ridge*, from which again small lateral passages are made, leading into every section of the mould. The general surface is then properly arranged with the aid of small *trowels*, and a little fine parting sand or brickdust is shaken over it. When this has been accomplished, the upper part of the flask is fitted to the lower by the pins, and then also rammed full of mould sand. The fine dry parting sand or brickdust serves to prevent the two halves from sticking together. A board is now placed on the top of the upper half, and struck smartly in different places with a mallet, after which the upper half and its board are lifted up very gently and quite level, and then turned over, so that the upper half rests inverted on its board. The models are next removed, and channels scooped out from the cavities left by them to the hollows or pouring holes (*ingates*) at the end of the flask. Solid *cores* of sand or metal are adjusted in the proper places when the article is required to be cast hollow (brass cocks, for instance), and also iron rails intended to have brass heads cast on them, or such other articles of iron as are required to be solidly united with the brass. The faces of both halves are now finally dusted with waste flour or meal dust; the two halves are then replaced upon each other, and the box is fixed together by screw clamps. The moulds for *fine castings* (articles with ornamental surfaces, as screens, sconces, bell-levers, &c.) are faced with various fine substances, such as charcoal, loamstone, rottenstone, &c. that they may retain a sharp impression; after which they are most carefully dried.

Braces. Set Moulds. Cores.

For ordinary work it is generally considered better that the sand should retain a little moisture, though great care must be taken in this respect, to guard against the danger of explosion.

The *mould* then is a square frame, mostly of iron, filled with peculiar dark red sand, which is pressed into a firm mass, in which the patterns of the casting are imbedded and their perfect shape impressed. The casters work at a large *trough* filled with the sand, and the workshop, with its forge, has some resemblance to a bakehouse where black bread is being kneaded into loaves. The first mould is made for what is called the "odd side" of the pattern—that is to say (in solid castings), the lower, or inferior side—and this serves as a sort of pattern to which the moulder refers in fine castings. The pattern being lifted off or out as soon as the sand-mould is sufficiently solid, the whole surface, in which the chasing of the pattern is clearly defined, is dusted with bean-floor or pounded "pot" first, and afterwards with loam, sand, charcoal, or coal-dust. This has the effect of making a smooth surface, and effectually filling the interstices in the sand, so as to prevent any raggedness in the casting. Each mould, or rather the two sides of the mould, are then placed near the surface and slightly baked, a channel having been made in the edge of each for conducting the melted metal to the pattern. The two sides are then placed together and held firmly by their pins and sockets, and the mould is ready for the casting. The "pots," or crucibles of greyish clay, which turn red by the action of the fire, are in the furnaces like so many tall flower-pots. The dirty yellowish brass ingots, made on the premises at a large mixing furnace, having been first placed across the tops of the pots, that they may expand before being melted, are about twenty minutes afterwards reduced to a molten mass, above which hovers a light sea-green flame mingled with streaks of brilliant colour, like the water from a dyehouse; meanwhile the moulds have been placed in a slanting position, with the opening in the side upwards, against a bank of sand or brickwork, and everything is prepared for pouring. A man, who should be strong in the wrist, stands on the *furnace*, which has the openings at the top, like a French cooking-stove, and taking off the brick covers from the square aperture, whence rushes out a tongue of green flame, lifts out the *pot* with a pair of *tongs*, and after the dross is removed by the *skimmer* or *grunter*, hands it to the pourer, who fills each mould in succession. The fumes which rise from the midst of the

Large Tongs. Small Tongs. Pot Holes. Skimmer. Grunter.

coloured fire are peculiar and penetrating, and the zinc eliminated from the molten brass falls in a metallic snow-storm, its flaky particles adhering to everything with which they come in contact, while the resistance of the sand to the metal causes a series of reports like muffled pistol-shots.

The brass cocks and plugs used in gas-fittings are all cast in one central stem, like cherries on a stick, their hollow forms being secured by means of *cores* made of hardened sand placed in the shape impressed in the mould. These are broken off the central stem with a pair of *pincers* immediately after casting.

The ornamental "vases" and larger ornaments which form the body of ordinary gas chandeliers and lamps are formed out of thin metal by a process called "stamping out," the plates of metal being placed on a hollow die, upon which a heavy hammer, or rather weight, is brought down, being released from a latch and worked by the foot. The depth of the casting would make so heavy a blow necessary that there would be danger of splitting the metal, an accident which is prevented by the introduction of a leaden shape and a layer of clay, which is decreased after each blow of the hammer until the proper depth is gradually secured without injury.

The process called "reversing" is an operation which secures a hollow casting, the inner or hollow side being called the "reverse." For this purpose a mould is made from one in wax, and the impression in the mould hardened, so that another model can be taken from it. This enables the moulder to secure a core which fits the impression in the mould, as one cup would stand inside another; and between the mould containing the sunk pattern and that with the projecting core there are placed strips of black clay, to secure sufficient thickness of metal, by not allowing the hollow to be too accurately filled. The pattern when cast is "laid out" on a hollow hemisphere of iron filled with pitch, and the irregularities of the casting removed by hand tools, *files, rasps,* and *knives.* In the case of figures, such as cupids, &c. forming ornaments for candelabra, the various limbs have often to be modelled in separate "cores," which are afterwards baked hard, and put together like a puzzle-map, imbedded in the sand of the mould previous to casting. This requires great skill to effect successfully,

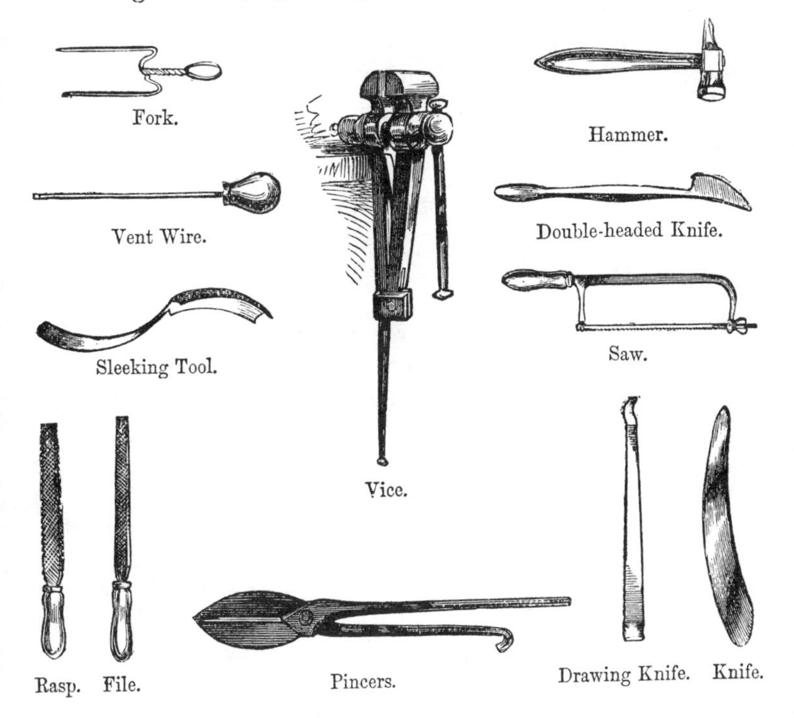

and an experienced "reverser" is a man of mark in the factory.

The completed castings are now removed to the chasing-room, where we may watch the gradual process of beautifying to which they are subjected, and the sharpening of the ornamental details by means of *tool* and *graver*, in a similar way to the first rough "laying-out," which removes the irregularities of the pattern. The arms and branches which form a part of the gas chandelier work, as well as many of the scroll-work ornaments, are cast in halves, which are taken to the soldering-room, where a workman, seated at a forge-like furnace, heats them in the burning embers, and applies to the edges the solder, with which is mingled a flux of borax and water to secure its melting. The heat is increased by a blowpipe, which is in reality a double or jacketed tube, the inner one supplying gas, and the outer being connected with a large pair of bellows, and mixing atmospheric air with the lighted gas at the point of combustion.

The pickling room is a large shed-like place filled with tubs, troughs, and earthen pans. Into one of these, containing diluted aqua-fortis, the metal is plunged for the purpose of removing the scale produced on the surface by the action of the fire; from this it is dipped in a stronger solution, to undergo the process called "fizzing," and its final baptism in pure acid restores the beautiful primrose colour which properly belongs to it. It is still dull, however, and goes to be "scratched," an operation effected by means of a revolving wire-brush, turned by a wheel and treadle, and kept continually wet with water.

The ornamental processes have next to be considered, and these are many. Previous to burnishing, the work is dipped in argot or tartar (the lees of wine-casks steeped in water), so that it may be subject to a strong antioxyde. The burnishing itself produces those bright veins and ornamental surfaces so often seen in brass work, and is effected by fixing the work in a vice, and rubbing the parts of the pattern which are to be brightened with a steel tool having a smooth bevel edge. After being treated with ox-gall, bean flour, and acid, to remove any still adhering grease, the work is dried, by being first dipped in hot water and afterwards buried in a pan of warm sawdust. Then there is lacquering, both black and white, a simple process enough, since the lacquer is laid on with a brush, and the work dried on a warm plate. Much of the work of the Brass-founder, as far as regards these latter operations, is of course effected by machinery, but the casting itself is entirely completed by the skill of the workmen.

THE GILDER.

A Gilder's Workshop.

WHEN once the house is built and the work of bricklayer, carpenter, plumber, painter, glazier, and mason is finished, it is necessary to set about those decorations which accompany the furnishing; and one of the first of the trades needed for this purpose is that of the Gilder, who has to do not only with cornices, mouldings and other ornaments, but also with the frames of pictures and looking-glasses that adorn the walls and chimney-pieces.

These frames, however, have first to be made by the joiner, and then receive the work of the carver, or the ornament maker. The joiner does little more than put the plain groundwork of the frame together; but the duty of the carver is of a very artistic description; and to be a good carver in wood requires an education and a taste very nearly equal to that of the sculptor, with whom the artists in wood formerly held a high rank.

Most of the ornaments now used for frames, however, are less expensive than those formerly produced by the carver who added a fresh value to the painting or the looking-glass by exercising his skill upon the costly settings in which they appeared. Composition ornaments are now in general demand for all but the most expensive frames, and as this composition—which is formed of glue, water, linseed oil, resin, and whiting—is pressed into moulds when it is of about the consistence of dough, it is evident that the mould maker has partially taken the carver's place. A new substance, however, has to a great extent superseded the old composition, and this is *papier-mâché*, or the pulp of paper (literally, mashed or beaten paper), which, from its lightness, its greater strength and durability, and the thinness to which ornaments made of it can be reduced, is preferable for all large decorations.

Whatever may be the size or pattern of the frame, however, we will suppose that the ornament maker received it from the joiner, who puts it together after it has been covered with coatings of hot size and whiting; the size being made from parchment cuttings or kid leather parings boiled to a sort of jelly. The nail or screw holes are then filled up with putty by the help of the putty knife, and the surface of the frame smoothed with pieces of pumice stone. The ornament maker next fixes on the decorations and hands it to the Gilder, whose first business is to wash the ornaments carefully in order to remove any oil that may have remained on their surface from the inside of the cast.

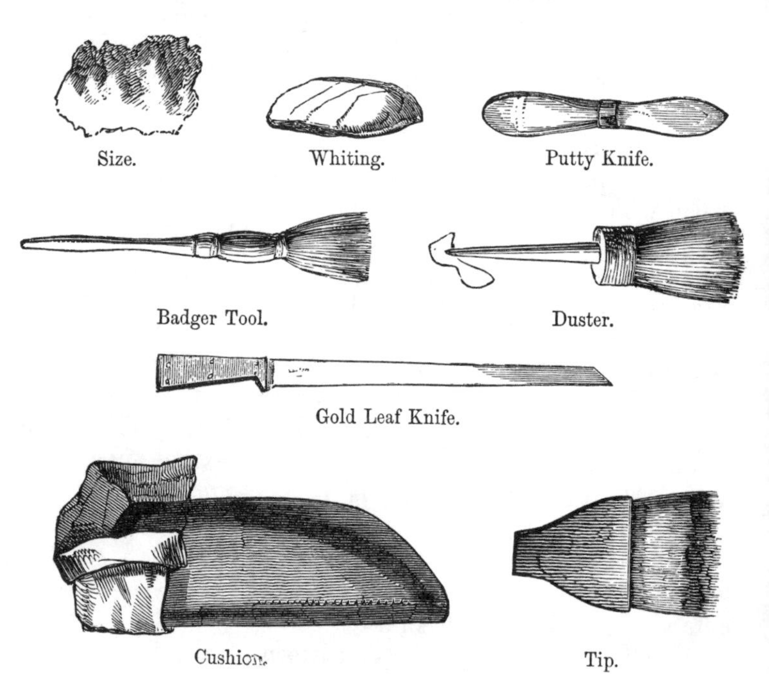

Size. Whiting. Putty Knife.

Badger Tool. Duster.

Gold Leaf Knife.

Cushion. Tip.

The principal tools required by the Gilder are, first: the *cushion*, which is a flat board covered with several layers of woollen or flannel and afterwards with a piece of leather, which is stretched tightly over it and nailed down to the edges, thus forming a firm but soft and elastic bed. A rim of parchment carried round one end of the cushion serves to hold the gold leaf. Second: a *gold leaf knife*—which is a straight smooth-edged instrument, not very sharp, but carefully pointed at the end. Third: the *tip*—a tool generally made of two pieces of card or very thin board glued together and holding between them a row of camel's hairs. It is also necessary to have a *badger tool* and a *duster;* the first for removing the loose edges and

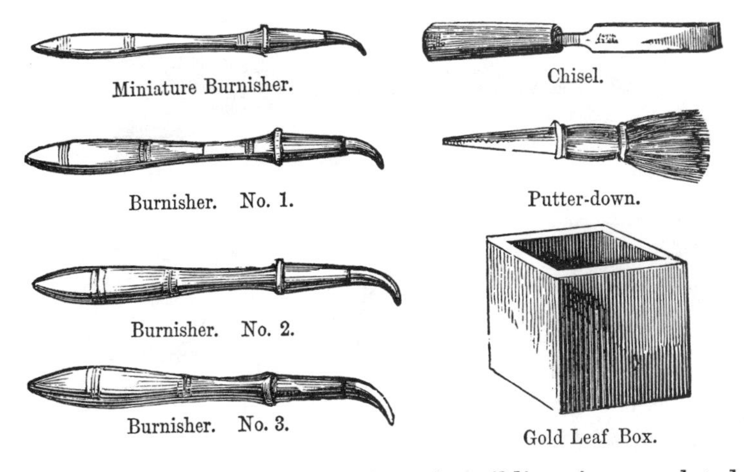

Miniature Burnisher. Chisel.

Burnisher. No. 1. Putter-down.

Burnisher. No. 2.

Burnisher. No. 3. Gold Leaf Box.

flying scraps of gold leaf, after the gilding is completed, and the latter for brushing away dust from the frame.

The *burnishers* are pieces of smooth stone (flint or agate) set in handles, and are used for rubbing some parts of the gold, when it is set on the frame, until it attains a brilliant polish and smoothness of surface: the bright gold portions of a frame being known as *burnish*, and the dull parts or "dead gold" as *matt*.

A *chisel* and *knife* are necessary for removing any inequalities or overflowings of size from the edge of the frame; the *feather duster* is used to dust the new gilding before it is sent home; the *size pot* is the vessel in which the size is melted, and the *pan* receives it before it is laid on. The *putter-down* is a large soft brush used for pressing down the gold into the ornaments, and removing the ragged edges of the gold-leaf from those parts of the frame where it is most difficult to place it smoothly.

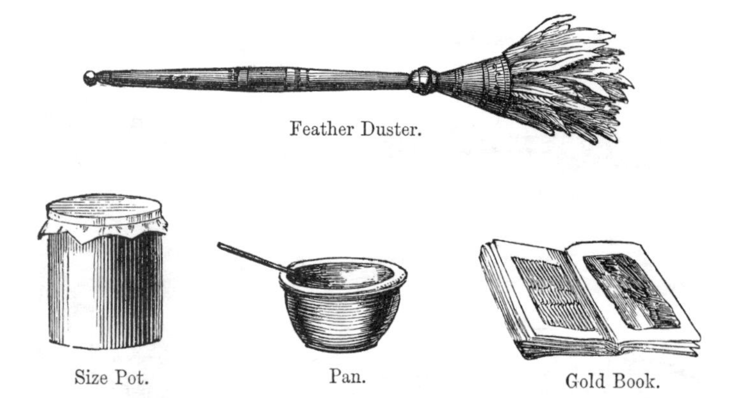

Feather Duster.

Size Pot. Pan. Gold Book.

The *gold box* is of course a receptacle for the gold leaf, and as the leaf is sent from the gold beaters in sheets placed between leaves of paper sewn together like a volume, this is naturally called the *gold book*.

Having received his frame, then, the Gilder first gives it two or three coatings of "thin white," which is the name by which the size and whiting is known in the trade. If any part of the frame is to be burnished, it afterwards receives a coat or two of rather thicker size.

When these are dry, strong warm size is laid on with a brush; this is called *clear cole*, and produces a smooth glossy surface, which prevents the *oil gold size* from sinking through. Oil gold size is the next coating given to the frame, and it is made of ochre and boiled linseed oil, ground up together into a smooth creamy liquid, which is thinned with more boiled linseed oil, and put on very carefully with a soft brush. In a few hours, after the oil gold size is put on, it is sufficiently dry to receive the gold leaf; the surface being then slightly sticky, so that it will hold it firmly and without its own surface being disturbed.

Pipkin.

Rule.

Gold Size Pot.

The Gilder now commences the most important part of his work by taking the cushion on his left hand, with his thumb through a loop which is attached to the bottom of the leather; between the fingers of the same hand he places the tools that he will have to use,—namely, the tip, the gold knife, and the camel-hair pencil. He then takes a gold book and carefully blows out a leaf at a time on to the cushion, until he has eight or ten leaves all heaped together within the rim of parchment which holds them from flying away. This is a very delicate operation, since if he should blow too hard the leaves would be carried all over the room. He next separates one of the leaves from the rest with his knife, and without cutting or tearing it lays it down smoothly upon the front part of the cushion, partly by a gentle use of the knife itself, and partly by skilfully blowing upon it. Then taking the tip in his right-hand he carefully presses it on the leaf, to which it adheres, and by this means transfers it to that part of the frame where he is at work. Where the ornaments are very deep, the same part is gilt three or four times over, and the gold is sometimes pressed in with a wad of cotton.

Skewing Brush.

Small Pencil.

After the whole surface is carefully *skewed*, or gone over with the brush which removes the ragged edges and still further presses the gold into the ornaments, the frame is well dusted with another soft brush, and then sized with clear size, after which the work is complete.

Supposing however that any part of the frame is to be of burnished gilt, the clear cole and the oil gold size must not be suffered to touch that portion,—burnish gold size, a substance made of grease, clay, black lead, red chalk, and bullock's blood, is used instead of the oil gold size. Several coats of this are laid on the part to be burnished, each being allowed to dry, and perfectly smoothed, before the other is applied. The surface is then washed with a sponge and clean water, another coating of gold size is laid on, the gold leaf is applied, and the burnishing tools used to impart the required lustre.

THE CABINET MAKER AND UPHOLSTERER.

CABINET MAKER'S SHOP.

THE trades of the Carpenter and the Joiner having been considered, we may now turn to that of the Cabinet Maker, who, though he makes the furniture of the house, and seldom has anything to do with building or fitting the house itself, uses many of the same tools as the joiner.

As the Cabinet Maker mostly works in more costly woods, and the operations of his trade have to be performed with greater nicety, his implements are generally of rather a better sort; while he has to fashion the articles in which he deals in so many different shapes that some of his tools, such as *planes* and *gouges*, are constructed especially for him, like the *panel plane*, used as its name

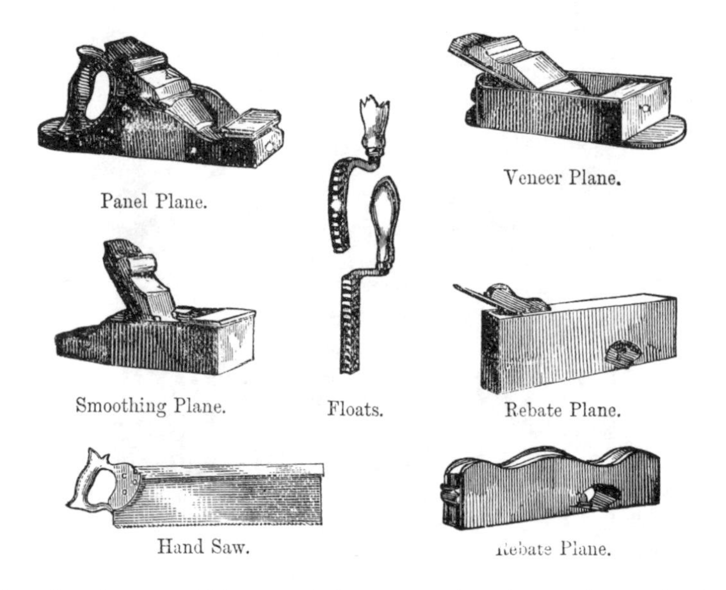

Panel Plane. Veneer Plane.

Smoothing Plane. Floats. Rebate Plane.

Hand Saw. Rebate Plane.

implies in smoothing and forming the edges of panels for wardrobes, chiffoniers, and other pieces of furniture before they are placed in their frames, and the *veneer plane*, intended for putting on *veneers*, or the thin slabs of costly wood with which more common woods are frequently covered. The *smoothing plane* and the *rebate plane*, as well as the *hand saw*, the *tenon saw*, the *gimlet*, and the *rule* and *square*, have already been mentioned in connexion with the joiner's business.

Then there are moulding planes, with their blades shaped hollow so that they will cut a strip of wood into a rounded form, or shaped round so as to cut a hollow groove.

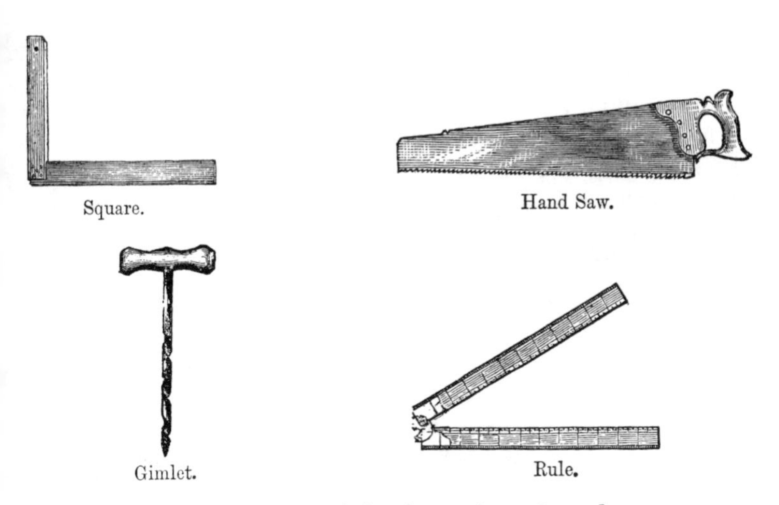

Square.

Hand Saw.

Gimlet.

Rule.

Sometimes boards are joined at the edges by a process called match boarding: a groove being made along the edge of one board while the edge of another is cut with tongue along the middle to fit accurately into the groove. For this purpose a pair of match planes are used, one of which makes the groove and the other the tongue exactly of the proper size to fit perfectly. This kind of joint is used for common doors, which it is not worth while to frame together in panels. The boards after being *matched* are nailed close together to strong cross-pieces.

The operations of mortising and dovetailing have been described in the description of joiner's work. The various fittings and joints used in making chairs, couches, tables, cabinets, side-boards, and other furniture are adaptations of the same kind, or differ only according to the shape and position of the various parts. In ornamental cabinet work the separate parts, such as pillars, legs, arms, and other pieces, are often supplied by the turner and the wood carver, who sometimes carry out their designs under the direction of the Cabinet Maker.

Mahogany and many other of the harder woods are difficult to work, as the grain does not all run the same way, so that in planing them the wood is likely to split or chip where it should be shaved off smoothly. To remedy this inconvenience, the Cabinet Maker's planes are furnished with double irons, that is, an iron with a flat dull edge is screwed on to the face of the cutting iron, so as to prevent the shavings chipping against the grain. The more cross-grained the wood is the closer the workman brings down the dull iron towards the edge of the sharp one, and his shavings are consequently finer.

The *veneering plane* is about the same size as the *smoothing plane*, but the iron instead of having a smooth edge is toothed like a fine saw, so that, instead of taking off shavings, it makes scratches all along the grain of the wood. This is applied to the veneer as well as to the wood to which the veneer is to be glued, so that the glue may easily hold the two rough surfaces together.

Previous to the veneer being put on, the work is well warmed before a fire, and the glue brush worked freely over both the veneer and wood to which it is to be applied. When the veneer is put on, it is rubbed backward and forward, at the same time being pressed down with the hands until it sticks in the right place. There are often lumps here and there where there is too much glue, and these are remedied by the *veneering hammer*, the head of which is made of wood furnished with a strip of iron plate. This strip is laid flat on the veneer, and the head of the hammer pressed with the hand while it is worked about by the handle, pressing out the glue as it moves towards the edge. When a piece of furniture is too large to be covered with one veneer, these thin slabs of wood are laid on in several pieces, the edges being first planed quite straight and made to meet with the greatest accuracy. The whole surface is afterwards worked with the toothing plane, and then scraped with a flat square piece of steel, which takes off a wonderfully fine thin shaving and leaves the surface perfectly smooth. It is afterwards finished with sandpaper.

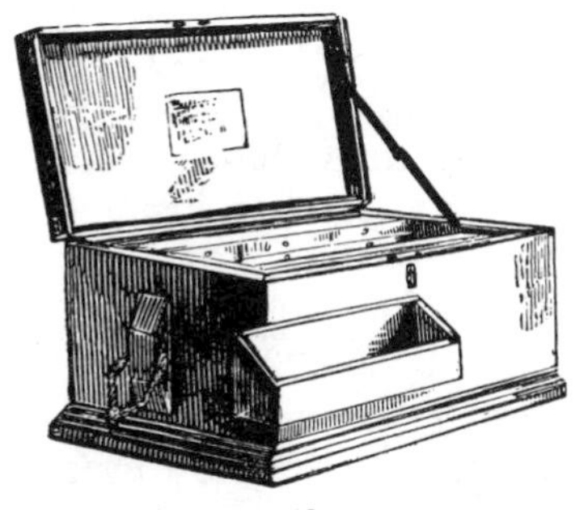

Tool Chest.

It is then *French polished* with a liquid composed mostly of rectified spirits of wine, gum, shellac, gum seed lac, and Venice turpentine, the furniture being previously well oiled that it may better receive this sort of varnish.

Clamps are a sort of screw vice for holding the various parts of the work. The *bow saw* is a small fine blade of steel notched like a saw, and fixed to a short handle, from which a wooden or metal bow extends to the other end of the blade. The bow keeps the saw from buckling or breaking, and the tool is used for small work, like the

Bow Saw.

Clamps.

fretwork in front of pianos, where a corner has to be turned and the piece sawn out. The *screws* seen in the larger picture hanging above the Cabinet Maker's bench are used for holding pieces together after being glued, or on other occasions.

It is supposed that there are about 50,000 workers in wood in London, and 350,000 in all England. About 160,000 timber trees of average size are required to make the furniture for the new houses built every year in England and Wales. In cabinet making there are many departments, such as the chair maker, the bedstead maker, the carver, the general manufacturer of tables, drawers, sideboards, wardrobes, &c. and the fancy Cabinet Maker, who uses costly woods and makes workboxes, desks, dressing-cases, and similar articles.

A good set of Cabinet Maker's tools is worth from £30 to £40.

The Upholsterer, whose trade is generally joined to that of the master Cabinet Maker, does what is called the "soft work," that is, he undertakes the curtains, hangings, cushions,

Upholsterer's Shop.

carpets, beds, and the stuffing of the seats of chairs. For these operations he requires but few tools.

The *devil* is the ugly and very absurd name given to a machine consisting of a box, inside which a spiked wheel turns; the use of this implement is to separate and tear to pieces such woven substances as, when reduced to shreds, serve for the stuffing of furniture, as also to soften and make finer hemp or tow for the same purpose, when horsehair, which is the best and most expensive material, is not used. Cocoa-nut fibre is now sometimes applied for this purpose; and in the commonest furniture hay is frequently placed as a foundation, with a small quantity of

Devil.

Bench Screw.

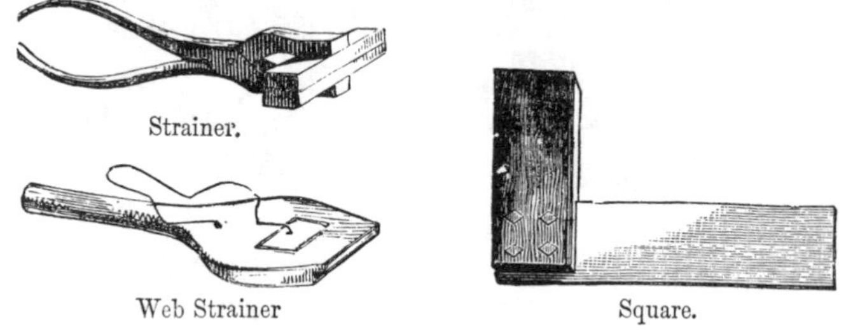

Strainer.

Web Strainer

Square.

horsehair on the top. The *bench screw* is a kind of vice which will hold a very thick substance, like the seat of a chair or sofa, without injuring the woodwork; the *web strainer* is used for stretching strong cross-pieces of webbing across the bottom of the seats of chairs or couches, to make a firm foundation for the stuffing to rest upon, and with the ordinary *strainer* to bring the canvas cover that confines the stuffing tightly and firmly to its place, an operation which requires great care, especially when metal springs are placed beneath the horsehair to make the seat more elastic. The Upholsterer's *hammer* is of such a shape

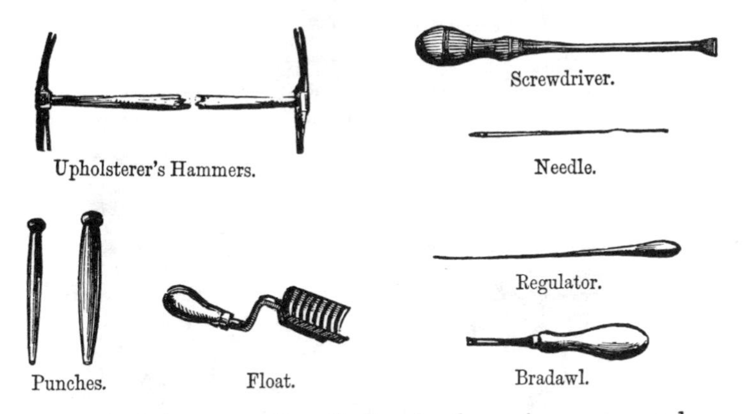

Upholsterer's Hammers. Screwdriver. Needle. Regulator. Punches. Float. Bradawl.

that it will drive a small nail deeply down in a space when it is hidden by the damask or leather covering of the furniture. The *punches* are for a similar purpose; while the *needle* and the *regulator* are used in stuffing the seats and properly adjusting the hair or other material.

FLOOR-CLOTH MANUFACTURER.

DRYING WAREHOUSE.

THE trade of manufacturing floor-cloth may be said to be connected with the furnishing of the house, since this very useful covering for the floors of halls and passages is now in almost universal use. Floor-cloth is generally made in large factories built for the purpose, since considerable space is required, not only for preparing, painting, and putting the pattern upon the cloth, but also for drying it when it is finished, the great lengths in which it is made rendering it necessary to hang it from a great height, in order that it may dry without the paint being damaged (*see drying warehouse*). The smell of the paint and other substances also makes it desirable to have the factory well ventilated, and situated at some distance from dwelling-houses.

The cloth is made partly of hemp and partly of flax, the former being the cheaper of the two, but the latter being fitted to retain the oil and paint on the surface without allowing it so easily to sink or soak through. In order to avoid the necessity for seams or joinings in the cloth, looms are constructed expressly for weaving canvas of the greatest width likely to be required. When the pieces of cloth are taken to the floor-cloth factories, they are generally either 100 yards long and 6 yards wide, 108 yards long and 7 yards wide, or 113 yards long and 8 yards wide. The flax and hemp are spun and the canvas woven principally in Scotland, in the town of Dundee.

Shears.

Cutting Knife.

The canvas is cut into pieces (*see cutting knife and shears*), varying from 60 to 100 feet long, and each of these pieces is stretched over a frame in a vertical position, most factories having a large number of such frames, some often 100 feet long by 18 or 20 feet high, and others of smaller dimensions. A wash of melted size is applied by means of a brush to each surface; and while this is wet the surface is well rubbed with a flat piece of pumice stone, by which the little irregularities of the canvas are worn down, and a foundation is laid for the oil and colour afterwards to be applied.

The preparation of japanned cottons, which are used for table covers, or what is known as "oil-cloth," is very similar.

The paint employed for floor-cloth consists of the same mineral colours as that used in house painting, and is mixed with linseed oil in the same way; but it is very

Blocks.

The finished Pattern.

Pattern for Floor Cloth.

Back of Printing Block.

much thicker and stiffer in consistence, and has very little turpentine added to it. The canvas receives several coatings on the back as well as on the front, and is well dried and smoothed at intervals.

The pattern is placed on by means of wooden blocks, on the first of which the rudiments, or ground work, of the design is cut, and on the rest other portions of the pattern,

Preparing Japanned Cottons.

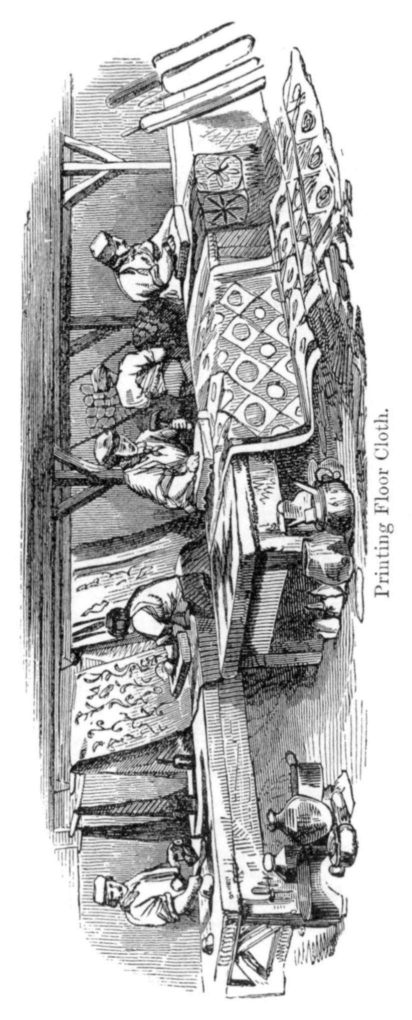

Printing Floor Cloth.

so that, as they are covered with paint and applied to the cloth by the workman, the pattern is gradually printed, and appears in the different colours which are successively applied to each block.

The blocks are made of pear-tree wood on one side and of deal on the other, the pear-tree wood being more easily engraved with the pattern.

The blocks (which we will suppose to be four for one pattern—red, yellow, blue, and green) being ready, and the prepared canvas spread out on a flat table, the printing commences.

The paint (say red) is applied with a brush to the surface of a pad or cushion formed of flannel covered with floor cloth; the block held by a handle at the back is placed face downwards on this cushion, and the layer of paint thus obtained on the surface of the block is printed on the canvas by pressing the block smartly down upon the surface of the latter. A second impression is made in the same way, by the side of and close to the first, until the whole surface of the canvas is printed over with the pattern of this first block, which is generally about 15 inches square. Then the second block is applied, and adds a little more to the pattern in another colour; the third follows, adding still more; and then the fourth, which completes the printing.

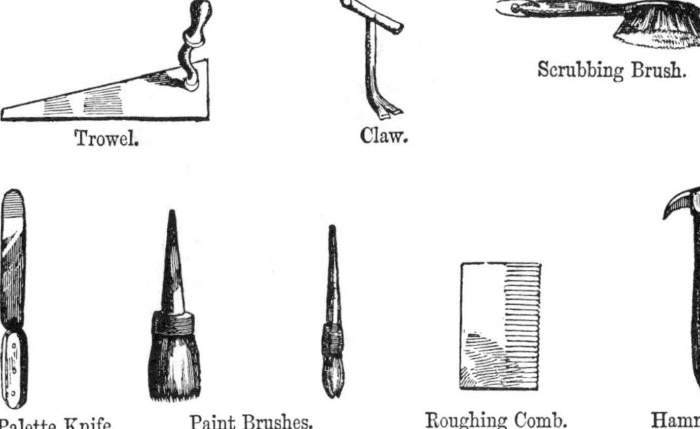

Trowel. Claw. Scrubbing Brush.

Palette Knife. Paint Brushes. Roughing Comb. Hammer.

The *trowel* and *palette knife* for spreading and mixing the paint, the *roughing comb* for patterns where the grain of wood is imitated, and the pots, cans, and jars, for the colours, are the principal tools besides those already described.

THE PAPER STAINER.

PRINTING PRESS.

THE trade of the Paper Stainer has grown to be one of considerable importance, and this is not to be wondered at when we consider how much the art of paper staining has increased the means of decorating our houses, by hanging the walls with elegant patterns printed in beautiful colours, instead of leaving them of one dull uniform hue, or a bare surface of wood and plaster.

In old times the walls of rooms were either of panelled wood, sometimes carved and polished, or were hung with tapestry made with the needle, or with woven silk, cotton, or linen, but the former was extremely costly, and the latter neither cleanly nor healthy. The trade of the Paper Stainer has to a great extent superseded both, and the interior walls of houses are now seldom formed entirely of wood, since they are intended to be covered with various qualities of paper hangings.

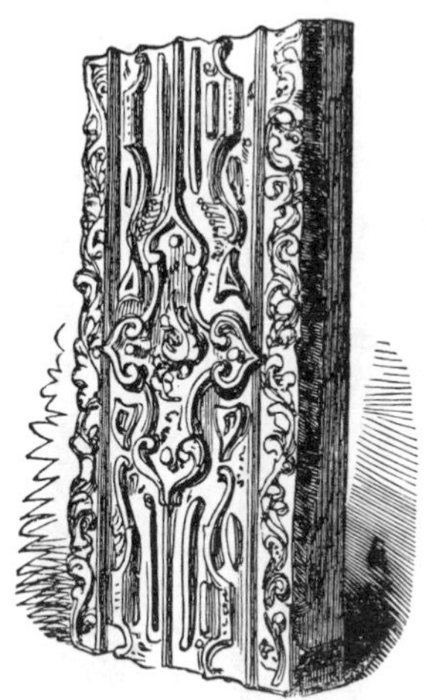

Front of Printing Block.

Back of Wooden Block for Printing.

The mode of printing or painting a pattern on large sheets of paper has now been in use for nearly two hundred years, although, of course, improved methods are at present employed.

There are three modes of producing the pattern on paper hangings. 1st. Wooden blocks are carved with the outlines of the figures only in relief; with these the paper is printed, and the pattern is afterwards finished by hand painting with a pencil. This mode is slow and too expensive for ordinary use. 2dly. A sheet of leather, tin, or copper, is cut with holes in the required pattern, and a brush dipped in colour is worked over the sheet after it is laid upon the paper, so that the paint goes through the holes, and leaves the pattern in colour. This is called stencilling, and is only employed for very common hangings. The third process

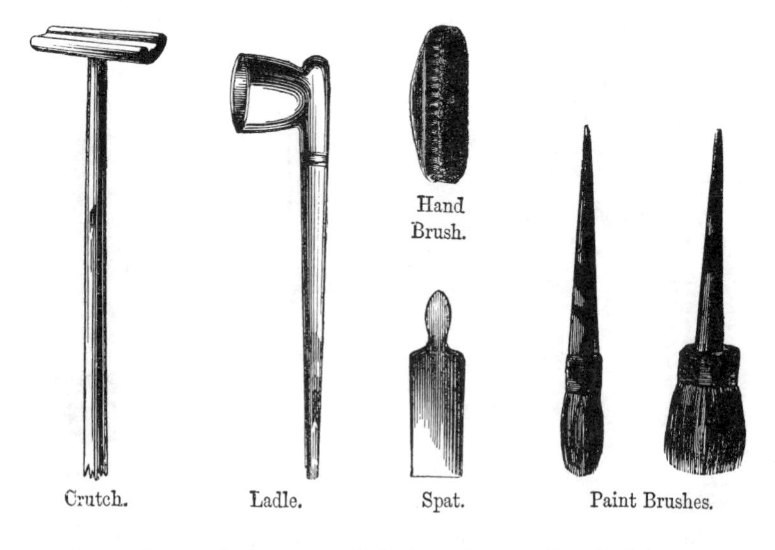

Crutch. Ladle. Spat. Paint Brushes.

consists of carving a wood block for each of the colours used in the pattern, and printing the paper by almost exactly the same method as that employed for printing floor-cloth, an operation which has already been described.

The paper is printed in pieces twelve yards long. A piece is laid out on a long bench and the ground colour applied, consisting of whiting tinted with some sort of pigment and liquefied with melted size. This is laid on with large brushes. When the paper is dry it is ready to receive the print at the *printing press*, where the blocks are pressed upon it by a sort of weighted arm which comes down from

Colour Sieve.

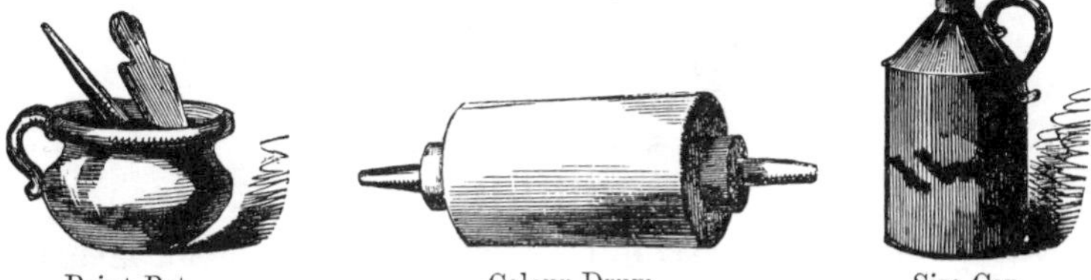

Paint Pot. Colour Drum. Size Can.

above the centre of the bench. There must generally be as many blocks as there are colours in the pattern.

Some paper hangings have a glossy or satin ground. To produce this a ground of satin white, properly tinted, is laid on; this ground is then rubbed with powdered French chalk, and worked with a brush till a gloss is produced. Sometimes these papers are passed between heated rollers which have been engraved with a sort of pattern, and this produces a pattern without any additional colour, like that of figured or watered silk.

Flock papers are those in which part of the pattern resembles cloth. To produce this the pattern is printed,

Drum for laying on Flock.

not in paint, but in size, and then the paper being passed through the *flock drum*, the flock (which is composed of fragments of woollen cloth) adheres to the pattern.

Striped hangings are sometimes produced by the paper being quickly passed on a roller beneath a trough, the colour in which flows through a number of parallel slits in the bottom; and occasionally various coloured stripes are obtained by dividing the trough into cells, with one cell and one slit for each colour. Some papers, in order to bear washing or cleaning, are printed with colours mixed with oil or varnish instead of size.

Paper Staining Machine.

THE CALICO PRINTER.

CYLINDER PRINTING MACHINE.

THE trade of the Calico Printer may be said to be one of the most important in this country, since we export such immense quantities of cotton prints to our various colonies, as well as to other countries in the world, that this business forms a very considerable part of British commerce.

Calicoes, muslins, &c. intended for printing must first have the fibres removed from their surface by the operation of the singeing machine; which consists of a half-cylinder of iron or copper laid horizontally, and kept at a bright heat

Bowking Kiers.

by a range of gas-flames or a furnace; over this half-cylinder the length of cloth is drawn with a steady motion till the down or fibre is singed off.

The next process is that of bleaching, because the whiter the cotton cloth becomes, the more light it will reflect from the surface, and the more brilliant the colour of the dyes will appear.

The principal chemical substance used for bleaching cotton is chloride of lime, which is known as bleaching powder; but there are several processes employed in its application, as well as various methods which are adopted by different manufacturers to increase its effects by mechanical contrivances, the application of heat, or otherwise.

Dash Wheel.

The first operation of bleaching, however, and that which immediately follows singeing, is boiling the cloth in an alkaline bath consisting of a solution of soda. For this purpose a *bowking apparatus* is used. This machine (*see cut*) is, in fact, a large cauldron, with a flat false bottom to protect the cloth from being scorched by the fire beneath. Through the centre of this false bottom a vertical pipe rises from near the real bottom to a height above the top of the cauldron, and carries a conical cap like an umbrella above its open end at top. When the liquid in the cauldron begins to boil thoroughly the steam forces a constant stream of liquid up the pipe, which stream is scattered by forcing against the umbrella-shaped cap, and so falls with some force back on to the cotton in the cauldron. When this process has continued long enough, the liquor is allowed to cool, and the cotton is taken to be rinsed at the *dash wheel*, where it is subjected to the free action of water, or to a

Wringing Machine.

rinsing machine, so constructed that the web travels on rollers, and is thoroughly washed during its course.

The simplest and earliest method of imprinting figures upon calico is by means of a wooden block, upon the face of which the design is cut in relief, as in an ordinary woodcut. The block is of sycamore, holly, or pear-tree wood, or more commonly of deal, faced with one of these woods. The block varies in size from nine to twelve inches long, and from four to seven inches broad, and it is furnished on the back with a strong handle. When the design is complicated, and a very distinct impression is required, the figure is sometimes formed by the insertion of narrow slips of flattened copper wire, the space between being filled with felt.

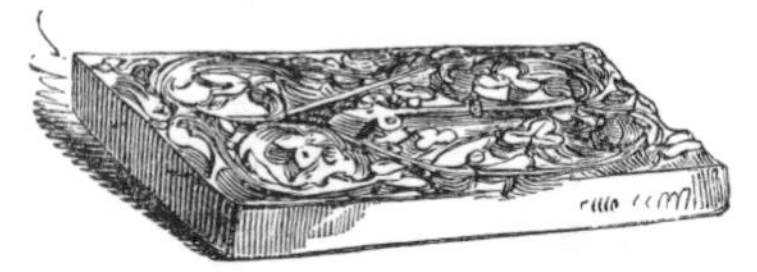

Face of Block for Calico Printing.

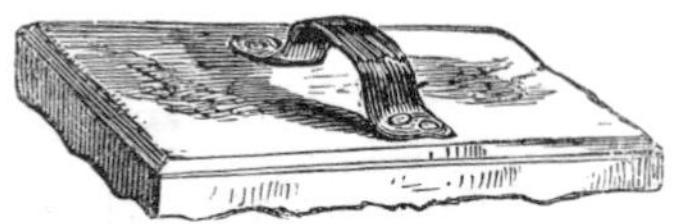

Back of Calico Block.

The printing block, which is worked by hand, is charged with colour by pressing it gently upon a piece of superfine woollen cloth called the *sieve*, stretched tightly over a wooden drum, which floats in a tub full of size or thick varnish to give it elasticity, so that every part of the raised device may acquire a sufficient coating of colour. The sieve is kept uniformly covered with the colouring matter by a boy or girl called the tearer, who takes up with a brush a small quantity of the colour contained in a small pot, and distributes it uniformly over the surface; for if this were not done, the block would take up the colour unequally.

The printing shop is a long well-lighted apartment, the air of which is kept warm for the purpose of drying the cloth as it is printed; to insure which it is passed over hanging rollers, so as to expose a large surface to the air.

The printing table, which is about six feet long, is made of some well seasoned hard wood, such as mahogany, or of marble, or flag-stone, the object being to present a perfectly flat hard surface. This table is covered with a blanket, upon which the calico is extended, and the block, being charged with colour, is applied to its surface, a blow being given with a wooden mallet to transfer the impression fully

Drying Room.

to the cloth. It is necessary, of course, to join the different parts of the design with precision, and in doing so the printer is guided by small pins at the corners of the block. Thus, by repeated applications of the block to the woollen cloth and to the calico alternately, the whole length of calico is printed.

By this method, a single block prints only one colour, so

that, if the design contain three or more colours, three or more blocks will be required, all of equal size, the raised parts in each corresponding with the depressed parts in all the others; in order, therefore, to print a piece of cloth twenty-eight yards long, and thirty inches broad, with three blocks, each measuring nine inches by five, no less than 672 applications of each, or 2,016 applications of the three blocks, are necessary. Thus it will be seen that printing by hand is a tedious operation, requiring more diligence than skill.

When the design, however, consists of straight parallel stripes of different colours, they may be applied by one block at a single impression. For this purpose the colours are contained in as many small tin troughs as there are colours to be printed. These troughs are arranged in a line, and a small portion of each colour is transferred from them to the woollen cloth by a kind of wire-brush. The

Brush.

colour is distributed evenly in stripes over the surface of the sieve by a wooden roller covered with woollen cloth. For the rainbow style, as a peculiar pattern is called, the colours are blended into one another at their edges by a brush or rubber.

An important improvement has been made in the construction of hand-blocks, by the application of a stereotype plate as the printing surface. A small mould is produced from a model of the pattern, and the stereotype copies are then made by pouring mixed metal into it. A number of the stereotype plates are then formed into a printing block, by being arranged in a stout piece of wood.

The greatest mechanical improvement in the art of calico printing was the invention of the *cylinder* or *roller*

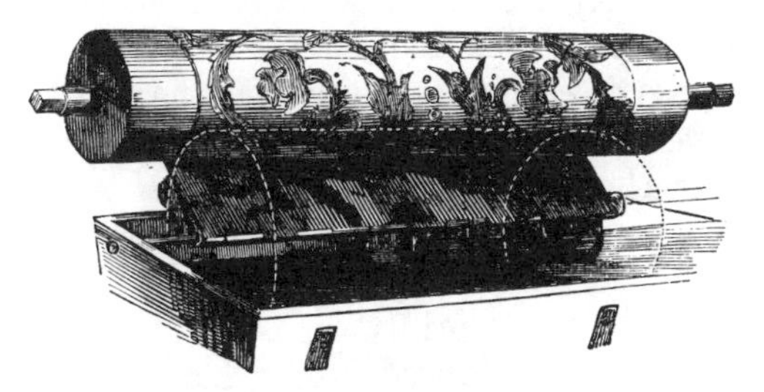

One of the Cylinders of the Machine, showing the way by which it is coloured.

printing at the end of the last century. This style of printing has been generally adopted in Lancashire, and is the cause of the success of the English over the continental printers. One cylinder machine, attended by one man to regulate the rollers, is capable of printing as many pieces as one hundred men and one hundred girls could print with hand blocks in the same time. A mile length of calico can be printed off with four different colours in a single hour.

This cylinder machine consists of a hollow cylinder or roller of copper, about three feet long and three or four inches in diameter, the pattern on which has been produced by the pressure of a mill, on which the design has been originally stamped by the pressure of a hard steel roller which has been engraved.

The copper cylinders are mounted on a strong iron shaft with a toothed wheel at its end, in order to put it in train with the rotatory printing machine for one, two, or more colours. On a roller at the upper part of this apparatus are wound the calico webs stitched together, the end of which is brought between the engraved copper cylinder and a large centre roller covered with blankets, against which it is made to bear with a regular pressure.

The engraved cylinder turns on the top of another cylinder covered with woollen cloth, which revolves at the same time as the former, while its under part dips in an oblong trough containing the dyeing matter, which is of a pasty consistence. The engraved cylinder is in this way supplied with plenty of printing colour, and is cleared from the superfluity by the thin edge of a blade made of bronze, called the *doctor*, which is applied to it as it turns, and gently scrapes the surface. After this the cylinder acts upon the calico, which receives the impression of the pattern in colour, and rolls onward at a great rate of speed.

There are various kinds of colours or dye stuffs used in calico printing, some of which impart fast colours by themselves, and others which require the web to be first prepared in order that they may become fixed.

In almost all the modes of calico printing the processes are very numerous to ensure the beauty and permanence of the colours. In what is called the *steam colour* printing, the agency of steam is applied to aid in fixing the colours to the cloth. The cloth is first steeped in a mordant or fixing liquor, then printed by the cylinder in various colours, called steam colours. It is then hung up to dry, and is afterwards exposed to the action of steam by means of various apparatus, which are adapted to the particular effect intended to be produced in fixing the dye.

The designs for calico printing are very expensive, and such a constant succession of new patterns are demanded, that some of the Lancashire printers expend several thousands a year on designing and engraving alone.

THE TINMAN.

WORKSHOP.

Tin is never found existing in an uncombined or native state. Tin ore occurs most abundantly in Cornwall and Devon, the mines of these counties having been celebrated for this metal from very ancient times. The district in Cornwall where tin mines are most abundant is termed the

"Stannaries." The Prince of Wales for the time being derives a large income from the mines, and is termed Lord Warden of the Stannaries. The word "Stannaries" is derived from the Latin *stannum*, tin. The amount of tin ore annually obtained from the Cornwall and Devon mines amounts to 11,000 tons, which, at the average value of £63 per ton, is worth £693,000. This ore yields about 7,000 tons of the metal, having an average value of £119 per ton. About four-fifths of all the tin raised in the world is produced in these mines. The ore is a heavy, hard, brittle, and usually dark brown mineral, which occurs chiefly in granular masses of various sizes. These grains are obtained in mines, where they occur in veins mixed with other minerals, and also from the beds of streams, where they have been washed out of the soil by the action of running water. In the former case they are termed tin stone; in the latter stream-tin.

Tin is obtained from the ore by first breaking up the latter, whilst a current of water flows over it and carries off the impurities, which are lighter, and therefore more readily borne away than the heavier tin ore. After being thus freed from the admixture of other minerals, the ore is usually roasted or burned, to drive off any traces of sulphur it may contain; it is afterwards heated to redness with blind coal or culm, and a small portion of lime, when the melted metal separates from the dross, and runs into cavities prepared for its reception. It is afterwards refined by being remelted, and other impurities separated, which either sink or float on the surface. The purest metal is yielded by the ore called stream-tin, which is smelted with charcoal instead of coal. If a block of tin thus obtained is heated slightly, and then allowed to fall from a height, it separates into a number of prisms which adhere together in pyramidal masses. In this state it is termed grain-tin. Tin is a silvery-white metal possessed of a high degree of metallic lustre. It is sufficiently soft to be cut with a knife, and may be readily bent, when it gives out a peculiar crackling noise; if repeatedly bent and straightened, it becomes hot from the friction of its particles with one another, and ultimately breaks. It is inelastic and moderately ductile, but very malleable, the thickness of tin-foil being about one thousandth part of an inch. It has but little tenacity, a wire of one-tenth of an inch in diameter not being able to support a heavier weight than 49 lbs.

Tin is the most easily melted of all the common metals, and it possesses the valuable property of not rusting when exposed at ordinary temperatures to the conjoined action of air and water, or even to weak vegetable acids. Tin, in a pure state, is seldom employed; but in combination with other metals it is a substance of great value. Its most important uses depend on its power of resisting the action of air and moisture; it is therefore largely employed for protecting the surfaces of copper and iron, that rust so readily.

Tin plate, or more properly tinned plate, which is so largely employed in the manufacture of saucepans, coffee-pots, tea-kettles, &c. by the tinman is not, as its ordinary name seems to imply, made of tin, but is formed of the best sheet iron, rolled out to the required thickness, and coated on each side with a layer of tin. Copper vessels ought to be invariably tinned inside, to prevent the rusting of the copper by the action of acids. Tinfoil is largely employed for the purpose of preserving moist articles from becoming dry, and is used instead of paper for enclosing fancy soap, chocolate, and other substances of a similar nature. It is also extensively used in the manufacture of looking-glasses, and is sometimes placed behind paper-hangings to exclude the damp of the walls. The most important alloys into the composition of which tin enters are bronze, pewter, bell-metal, and solder. Tin dissolved in acids is largely employed by dyers in fixing or rendering permanent various colours used in dyeing. The preparation termed putty powder is a rust of tin, obtained by exposing the melted metal to the air; it is employed in polishing metals and other articles.

The tinning of the inner surfaces of cooking utensils and other vessels of capacity is effected by scouring the surface

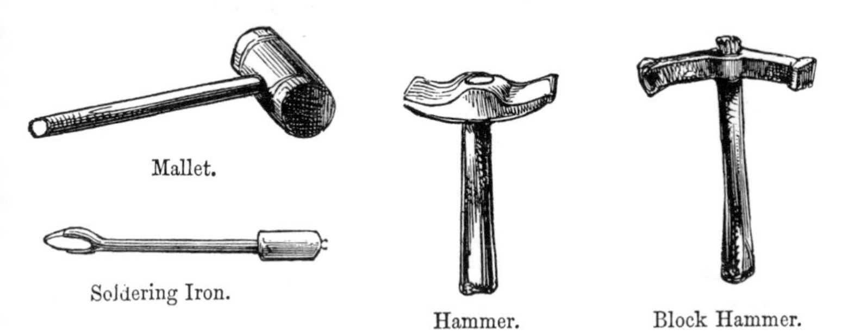

Mallet. Soldering Iron. Hammer. Block Hammer.

until it is perfectly bright and clean; then heating the vessel, pouring in some melted tin and rolling it about, and rubbing the tin all over the surface with a piece of cloth or a handful of tow; powdered rosin is used to prevent the formation of oxide. Bridle bits, stirrups, and many other small articles, are tinned by immersing them in fluid tin. Tin-plate working, or the forming sheets of tinned iron into a variety of useful vessels and utensils, is carried on by means of bench and hand *shears*, *mallets*, and *hammers*, *steel heads* and *wooden blocks*, *soldering irons* and *swages*. In the formation of a vessel, the first operation is to cut the plate to the proper size and form with shears, and, when the dimensions of the article require it, to join them together, which is done either by simply laying the edge of one plate over that of the other, and then soldering them together, or by folding the edges together with laps and then soldering them. Similar joints are required when gores or other pieces are to be inserted, and also at the junction by which

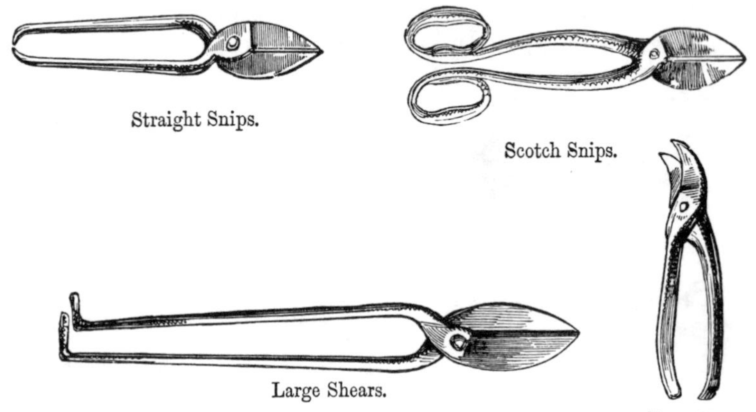

Straight Snips.

Scotch Snips.

Large Shears.

Bent Snips.

a cylinder is closed in. The usual method of forming laps, bends, or folds, for this or other purposes, is to lay the plate over the edge of the bench and to bend it by repeated strokes with a hammer; but a machine is sometimes used for this purpose.

After a tin vessel has been rounded upon a block or

mandril by striking it with a wooden mallet, and the seams finished, all its exterior edges are strengthened by bending a thick iron wire into the proper form, applying it to what would otherwise be the raw edges of the metal, and dexterously folding them over it with a hammer.

A superior kind of tin ware, commonly known as block tin ware, is carefully finished by beating or planishing with

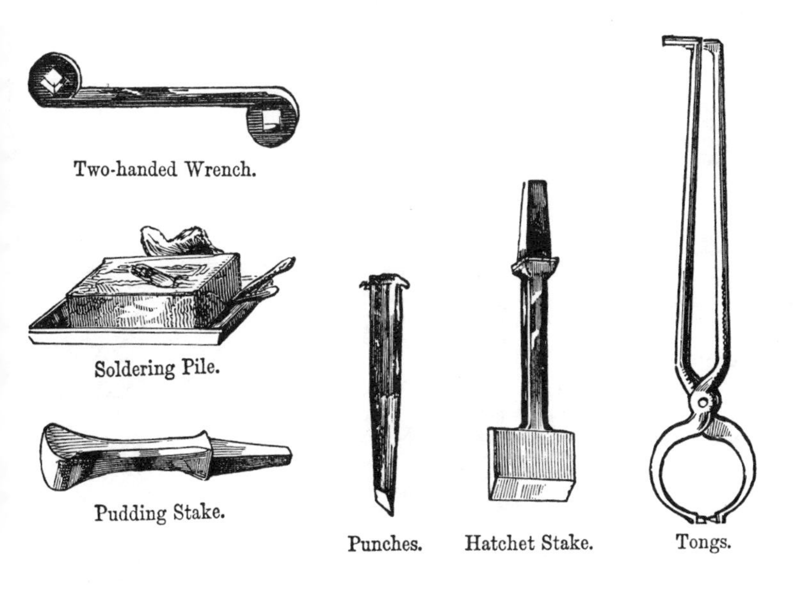

Two-handed Wrench.

Soldering Pile.

Pudding Stake.

Punches. Hatchet Stake. Tongs.

a polished steel hammer upon a *metal stake.* The process of swaging is resorted to as a ready means of producing grooved or ridged borders, or other embossed ornaments. This process consists in striking the metal between two steel dies or swages, the faces of which bear the desired pattern, and are made counterparts of each other. Many ornamental articles are produced by embossing or stamping tin plate, in the same manner as other metallic sheets, with a fly-press or other machinery. Cheap coffin-plates are manufactured at Birmingham in this way; and these and similar articles are sometimes lacquered, painted, or japanned. Tin forms the principal ingredients in various kinds of pewter and other white-metal alloys, which are manufactured into domestic utensils by casting, stamping, and other processes.

Chisels. Charcoal Stove. Polishing Anvil.

Britannia metal is a mixture of tin, antimony, copper, and brass, which is melted, cast into slabs, and rolled into sheets. The principal use of this metal is for candlesticks, teapots, coffee-biggins, and other vessels for containing liquids. The feet of candlesticks, the bodies of teapots, and other articles containing embossed work, are stamped between dies; while articles of a more globular shape are stamped in two or more pieces, and afterwards soldered together. The sheet metal has a ductility which enables

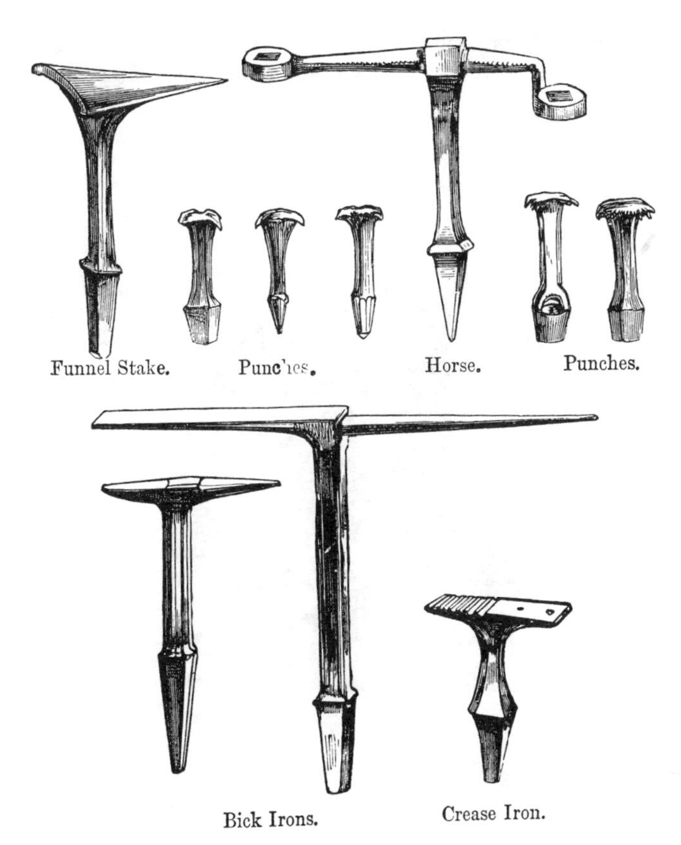

Funnel Stake. Punches. Horse. Punches.

Bick Irons. Crease Iron.

it to be bent into various curved forms by pressure on a model or core: this process is called *spinning*.

Many small vessels, spoons, and other articles, are cast in an alloy somewhat harder than that which is rolled into sheets. Most of the tools employed by the Tinman are the *irons, stakes,* and *bickers,* on which the tin is hammered into proper shape, with *shears* for cutting, the *punches* for piercing holes, and the *soldering iron* and *charcoal stove* for making joints.

THE FARRIER.

FARRIER'S SHED.

WHEN we remember how usefully horses are employed for our advantage, how generously and willingly they work, and how docile and obedient they are when properly treated, we shall begin to see that the trade of the Farrier is one which should be studied very carefully, and that nobody should follow the business who has not become tolerably skilful. The Farrier who shoes the horses, is very often consulted when those animals are ill, so that he should have some knowledge of simple remedies in cases when the veterinary surgeon lives at a distance, or is out of the way. Especially the Farrier should thoroughly understand the construction of the horse's hoof, which, hard and simple as it may look, is very delicate, and is composed of several important parts.

One thing should never be forgotten in shoeing a horse,—first that, although the hoof is a hard horny covering, it has an inside portion which is very tender and liable to be hurt;

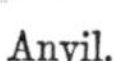

Anvil.

Stool.

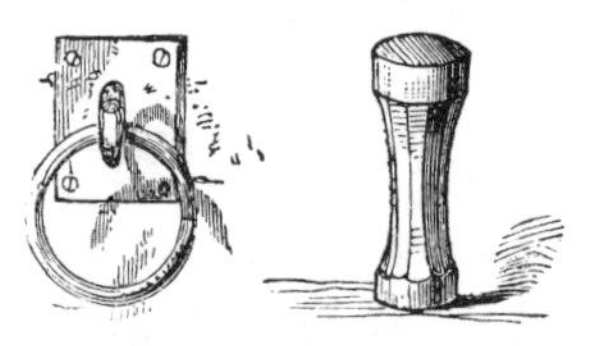

Staple. Pointing Stake.

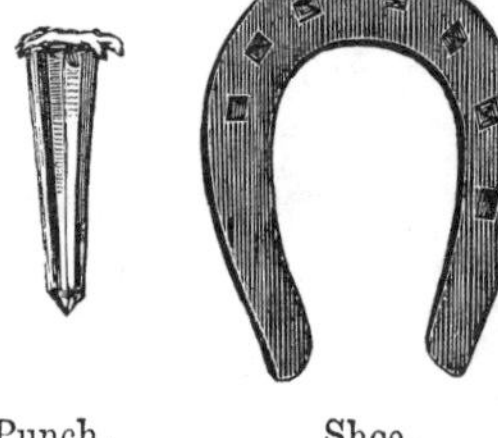

Punch. Shoe.

and secondly, that the hoof itself expands as the weight of the horse presses upon it.

The Farrier's shed is fitted with a forge, or furnace where the iron is heated, and in which the fire is blown to a great heat by the huge bellows fastened above it; it also contains an *anvil,* on which the *horseshoes* are made or shaped, a *stool* on which the Farrier sometimes sits to examine a horse's hoof, and *staples* and *rings,* to which the horses' heads are fastened by halters during the process of shoeing. When a horse is taken to be shod, the Farrier should begin by taking off one of the old shoes. He first raises the clenches with a tool called the *buffer*, and if the shoe does not then come off easily, loosens some of the

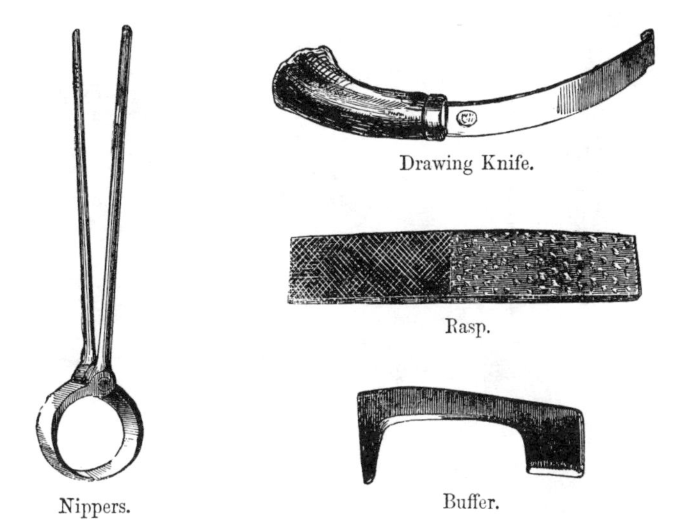

Drawing Knife.

Rasp.

Nippers.

Buffer.

nails with the *punch,* till it can be gently removed. When the shoe is off he rasps the edge of the hoof all round, and with the *nippers* or pincers takes out any stubs that may be left in the hoof. He then pares the hard portions of the foot, and this is an operation which requires great care and skill as well as a good deal of practice in the use of the *drawing knife.* The Farrier must always remember the state of the roads when he is paring the horse's feet, for if the roads are dry and stony he must take off very little of the horn, or the foot will be bruised.

The horseshoes are frequently purchased by the Farrier

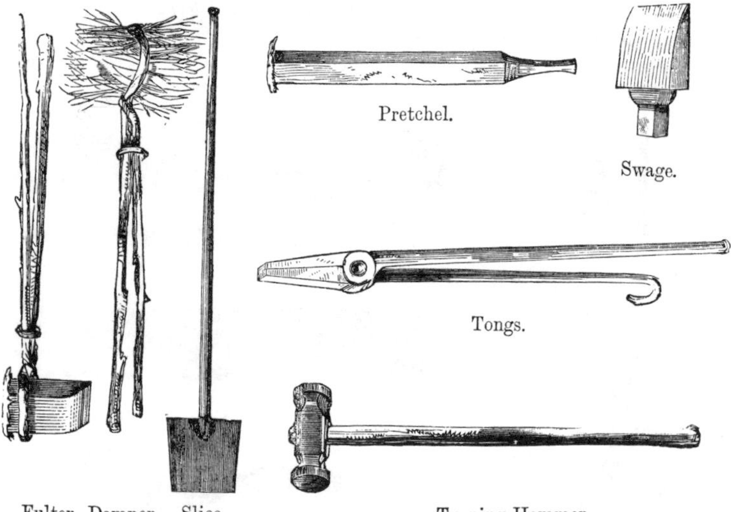

of the Blacksmith who makes them, but some Farriers are also Smiths, and both make and fit the horseshoes. In either case the Farrier keeps a stock of rough shoes which he alters at the time that they are wanted, so that they may fit the horse, and one of the first things to be done is to make the groove all round the shoe, and drill the holes in it for the nails. This groove, in which the heads of the nails sink, is called the "fuller," and the tool with which it is made is also called the "*fuller*," or "*fulter*." Having cut off the ends or heels of the shoe, made the fuller, and opened the nail holes, the Farrier next makes what is called the "clip," which means turning up the toe of the shoe, to prevent its being forced back on the hoof. In these parts of his work he has probably used the *chisel*, *turning hammer*, *swage* and *vretchel*, while for the work at the forge he has had to employ the *poker*, the *tongs* for holding the shoe on the *anvil*, the *slice* for taking small things from the fire, and the *damper*, which is a wisp of wet straw held by wooden tongs for lessening the heat of the shoe during hammering. He next begins to fit the shoe, the horse being tied up to the staple in the wall of the shed. The fitting of the shoe is an operation requiring the greatest care and attention, and the good Farrier will spare no pains to do his work perfectly, as many a valuable horse has been ruined by an ill-fitting shoe. When the shoe is fitted it is "filed up," by which all roughness is removed from the edges of the nail holes, and the sharp edges of the shoe itself are taken off.

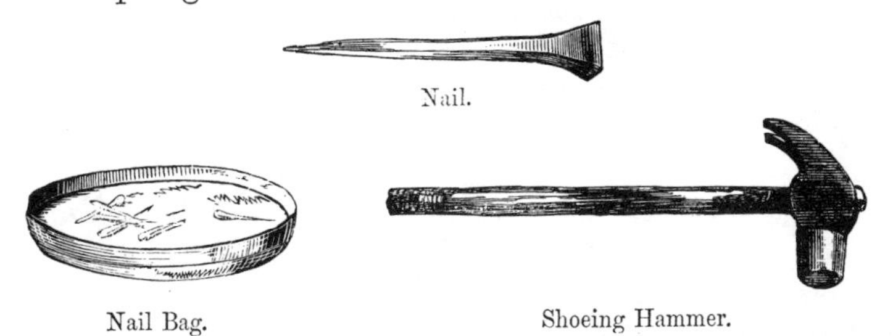

Nail.

Nail Bag.

Shoeing Hammer.

The Farrier generally makes his own *nails*, since they are of a peculiar shape, and the heads should completely fill the nail holes, that they may not allow the shoe to shift on the horse's hoof. They are made from long rods of iron called nail rods, and when finished are spread about the smithy to cool, because when they are allowed to cool gradually they become harder, and less liable to break.

If the nails are of a proper shape, the holes straight through the shoe, and the shoe fits the foot, very little skill is required to nail it on, and clench the ends of the nails to the hoof. Before the shoe is nailed on, however, it is usual, when the horse has tender feet, to cover the sole of the foot with leather, gutta-percha, or felt made waterproof, "felt"

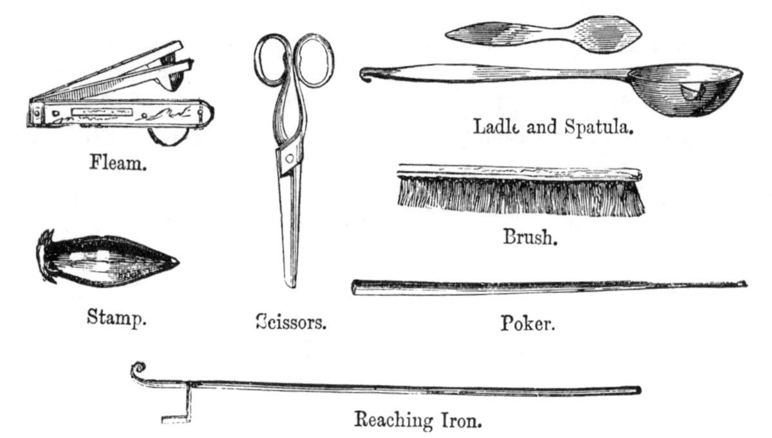

Fleam. Ladle and Spatula. Brush. Stamp. Scissors. Poker. Reaching Iron.

being made of woollen cloth torn to shreds which are then pressed together and formed into sheets. In any case the Farrier is expected to "stop the foot," that is, to fill the hollow and tender portion of the foot within the shoe, with tow or oakum dipped in tar.

The shoe is now nailed on, and a good Farrier will often be able to secure it with only five nails.

The hind shoes are of course different in shape to those which are placed on the fore feet, and it is generally necessary to use seven nails to fix them on to the hoofs, since the hind foot expands less than the fore, and there is more drag upon it when the horse is in motion, so that the shoe is more easily shifted.

The time at which a horse's shoes want removing depends on several causes. If a horse wear out his shoes in less than a month they had better not be removed, but whether the shoes are worn or not the horse's hoofs should be looked to by the Farrier, every three or four weeks, as the hoofs sometimes outgrow the shoes, and the shoes require refitting.

From what has been said about the Farrier's business it will be seen that it is a most important one, but besides the knowledge and experience required for the more mechanical part of his work, he should also know something of the diseases and ailments of horses, and be able to apply the proper remedies.

It is true that this part of the business belongs properly to the veterinary surgeon, but the Farrier should at all events understand what is proper to be done in any ordinary disorder or in cases of emergency. He will of course know how to use the *fleam* when the animals require bleeding, as they frequently do. This instrument is a sort of knife, the sharp part of which is the small spade-shaped pieces at the ends of the blades. The *ladle* for melting the ingredients of ointment, for sprains or swellings, and the *spatula* for mixing it, or for spreading and mixing the drugs for boluses or the large pills frequently given to horses, are some of the instruments used in this part of the Farrier's business.

THE NEEDLE MAKER.

GRINDING NEEDLE POINTS.

THERE is perhaps no implement of greater importance than that smallest of all tools, the needle, and in all civilized countries the number of needles consumed is so great, and such an enormous supply is required for the sewing of the clothes of mankind, that the manufacture is one of the most remarkable in this country, whence by far the greater part of the whole supply is derived.

It will only be proper in this place for the author to acknowledge his obligations to Mr. W. B. Tegetmeier, whose most useful little work on "Common Objects" affords brief and reliable information on this as well as on many other interesting subjects.

The material from which needles are made is soft steel wire of the requisite degree of fineness. This is obtained from the manufacturer in large coils, each containing sufficient wire to form several thousand needles. These coils are first cut up into pieces of the length required to make two needles, usually about three inches, large *shears* being used, capable of cutting a coil of one hundred wires.

Shears.

Soft Straight Liner.

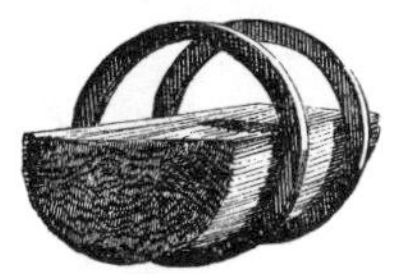

Needles placed in iron rings.

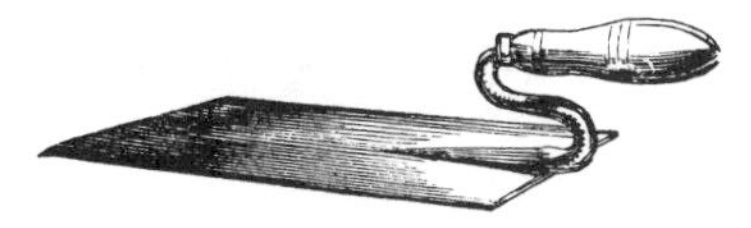

Trowel for hardening.

Five or six thousands of these lengths are made into a bundle kept together by a ring of steel at each end. They are then heated to redness in a furnace, and afterwards laid upon a flat iron plate, and rubbed backwards and forwards with a steel bar until each wire is perfectly straight.

The next stage is to grind a point at each end of the wire. This is done by the aid of grindstones about eighteen inches in diameter and four inches thick; they are made to revolve so rapidly that they are liable to fly into pieces, and are therefore partially enclosed in iron plates to avoid injury to the grinder, should such an accident occur. The grinder takes from fifty to sixty wires between the thumb and forefinger of his right hand; and

Rubbing.

as he presses them against the stone, he causes all the wires to roll round, and thus each is ground to a point. So expert do the grinders become by practice, that they point a handful of these wires, usually about sixty, in half a minute, or about seven thousand in an hour. During the grinding every wire gives out a stream of sparks, and these together form a bright glare of light.

Pointing these wires is the most unhealthy part of the manufacture; the fine dust is carried into the lungs of the workmen, and destroys them in a few years, very few living beyond the age of forty. Wet grindstones cannot be used, as the points of the needles would be rapidly rusted.

The wires thus pointed at each end are stamped by a heavy hammer, raised by a lever moved by the workman's foot. The under surface of this hammer is so formed, that

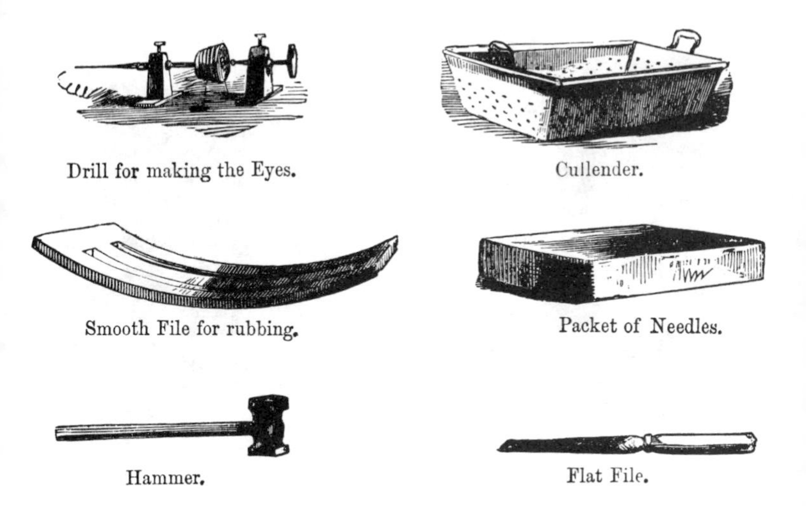

Drill for making the Eyes.

Cullender.

Smooth File for rubbing.

Packet of Needles.

Hammer.

Flat File.

when it falls on the wire midway between the two ends it stamps on one side the gutters, or grooves, in which the eye is afterwards made; and the anvil on which the wire rests when the hammer strikes it forms the two grooves on the opposite side. This stamping also makes a slight depression or pit on each side at the spot intended for the eye. The wires are then passed to a boy, who takes a number of them in his left hand, whilst with his right he works a press, moving two hard steel points or piercers. These come down upon the wire as it is placed beneath them, and pierce the eyes for the two needles. Each wire now resembles two rough unpolished needles united together by their heads; and as it would require much trouble to divide them separately into two needles, a number are threaded upon two very thin wires, and are separated by filing and bending.

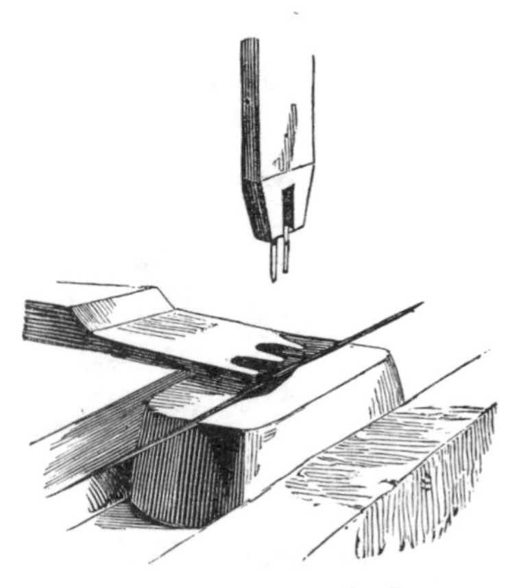
Hand Press for piercing the Eyes.

Stamping.

Any needles which may have been bent in the several processes are straightened by rolling under a steel bar, and are hardened by heating in a furnace, and suddenly cooled in cold water or oil. After hardening they are tempered by being slightly heated, and if any are bent during hardening, they are straightened by being hammered on anvils with small hammers; finally, the whole are polished by laying twenty or thirty thousand side by side upon a piece of thick canvas, smearing them with oil and emery, rolling up the canvas, and rubbing them under a press for several hours or even days.

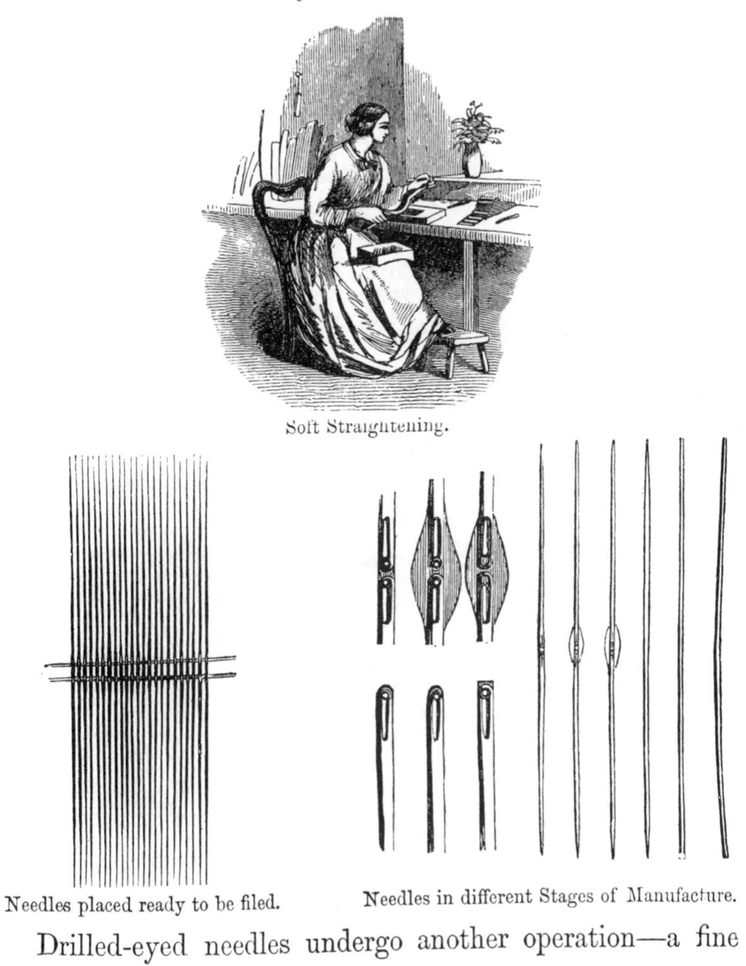

Soft Straightening.

Needles placed ready to be filed.

Needles in different Stages of Manufacture.

Drilled-eyed needles undergo another operation—a fine

drill is made to revolve rapidly in the eye of each, to take off the rough edge and to prevent their cutting the thread when used; finally, the points are finished on a revolving stone, and polished on a wheel covered with leather, and enclosed in a paper for sale.

Simple as the construction of a needle may appear, it has to pass through the hands of 120 workmen, from the time it leaves the iron mine until the manufacture is completed.

The chief seat of the needle manufacture in this country is Redditch, in Worcestershire, where upwards of seventy millions are made weekly. English needles are far superior to those of foreign manufacture.

THE CALENDERER AND HOTPRESSER.

THE business of the Calenderer and Hotpresser in so many respects resembles part of that already described as preceding the printing of calico, that only a brief notice of it will be necessary. The singeing and bleaching of the cotton has been explained, and calendering is the name generally applied in the manufacturing districts to the processes of smoothing, dressing, and glazing cotton and linen goods; the object being either to prepare them for the operations of the calico printer, or to impart the last finish to the goods before they are folded and packed for the market.

The earlier calenders, or calendering machines, closely resembled a common mangle in their action, but were very

Singeing Cotton Goods before Calendering.

Windlass.

Iron Plate.

Scissors.

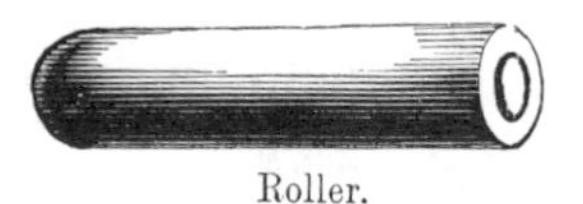

Roller.

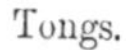

Tongs.

large and heavy, and worked by a horse-wheel or other sufficient power, but the process was greatly improved by the invention of a machine in which the pressure is produced between rollers, instead of between rollers and flat surfaces, and in which consequently the alternating movement is got rid of, and also it is easier to give a uniform and equal pressure.

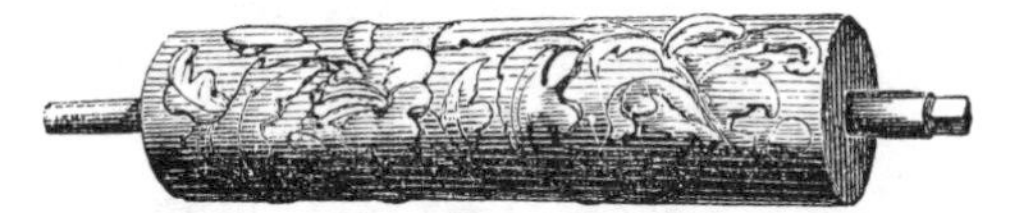

Embossed Roller for Machine Printing.

Presses.

The rollers or cylinders were formerly made of wood; they are now usually of paper or cast iron. The paper cylinders are formed by packing a great number of circular pieces of stout pasteboard upon an iron axis, and compressing them very tightly by means of iron bolts passed through them, acting upon circular end-plates of cast iron. The surface is brought to a perfectly even and polished state by turning in a lathe. Iron rollers are made hollow and when necessary heated from the inside. When a glazed or

Printing Press.

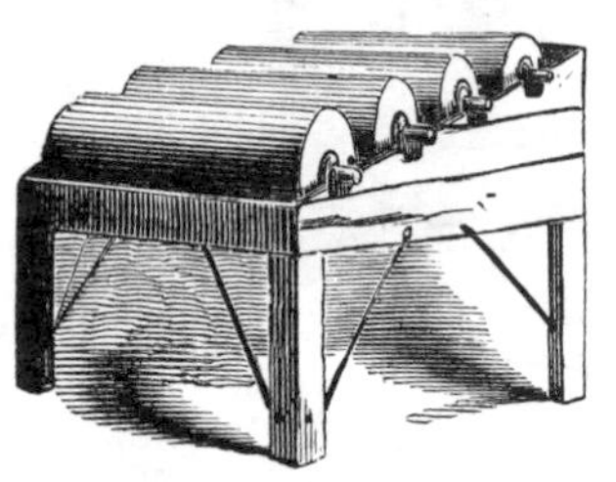

Chair with Rollers.

Tin Block.

polished surface is required on the goods to be calendered, mechanism is employed to cause two adjacent rollers to revolve with different velocities, so as to produce a rubbing action.

The *furnace* for heating the *rollers,* the *chair with rollers,* the *presses* and *tin blocks* for hotpressing, and the *windlass* for turning, are easily understood in relation to this trade. The *printing press* and *embossed cylinder* have been mentioned in the trade of calico printing. The *mill,* turned by a horse, is used where steam power is not employed to put the machinery in motion.

Furnace.

THE CUTLER AND FILE-CUTTER.

GRINDING.

AMONGST all the trades that occupy the attention of mankind, that of the Cutler, which also includes the tool maker, is certainly one of the most essential, since without tools no other manufactures could be carried on. Cutting instruments of various kinds have been in use from the earliest ages, if for no other purposes, for cutting food, slaughtering animals, and making war upon each other. In ancient times, as well as amongst some barbarous tribes at the present day, these implements were frequently made of shells, edged flints, or hardened wood, fashioned into sharp weapons; at a late period, cutting as well as warlike instruments were formed of brass or bronze; but at the present time, in all civilized nations they are formed exclusively of steel or iron.

Steel is formed from the purest bar iron—that which comes from the Swedish mines being preferred. This is buried in powdered charcoal and heated to whiteness for several days, without exposure to the air; during this time the metal becomes much harder, whiter in colour, crystalline in texture, and blistered on the surface. The blistered steel so produced is prepared for use, either by binding several bars together and hammering them into one, or by melting them in earthenware pots, called crucibles, and pouring the melted metal into moulds of the size required. In the latter state it is called cast steel.

Cutlery is generally understood to comprise all kinds of knives, razors, lancets, and edge tools, including scythes, saws, scissors, shears, spades, and many others; and the manufacture of forks, files, and some other instruments not possessing cutting edges, is frequently included in the business. It will be impossible to give a detailed description of how all these are made, so only two or three must be selected. In a Cutler's factory knife blades are forged from steel bars in a number of small rooms, each containing a fireplace or hearth, a trough to hold water, and another trough for coke, which is specially prepared for this kind of work; there are also an *anvil, hammers,* and some other tools.

Two persons are engaged in each room, one being called the maker or forger, the other the striker. The forger buries the end of the steel bar in the fire to the extent required; and to determine when it should be removed requires some judgment, since if it be overheated or "burnt," it will be quite unfit for cutting purposes. On the other hand, it must be sufficiently heated to acquire the proper

Cutler's Hammers.

Anvil.

degree of softness for the operation of shaping the blade from it. When the end of the bar has been properly heated it is brought to the anvil, where it is fashioned by the striker into the required shape by means of a few blows of the hammer. This roughly shaped blade is then cut off from the end of the bar, which is again heated for forming the next shape, and so on to the end.

The cutting part of the blade thus rudely formed is next welded to a piece of iron, which forms the bolster, or *shoulder*, that is, the part that rises round the handle of the knife. To make the shoulder of the size and shape required, and to give it neatness and finish, it is introduced into a *die* by the side of the anvil, and a *swage* (*see Blacksmith*) placed upon it, to which a few smart blows in the proper direction are given by the striker.

Shoulder-iron.

The die and swage are called *prints* by the workpeople. Besides the bolster, the part which fastens into the handle, technically termed the *tang*, is also shaped from the piece of iron welded on to the cutting part of the blade. After the bolster and tang have been properly finished, the blade is heated again, and then well hammered on the anvil. This operation, which is termed *smithing*, requires particular care and attention. It is intended to consolidate the steel, and to render it brighter. The next process the blade has to undergo is that of *marking*. This is done with a broad punch made of the very best and hardest steel, and having the name and corporate or trade mark of the firm carved on the bottom end or point. The blade is heated to a dull red (worm-red, as it is termed by the workmen), and the mark cut in on one side of the blade with the punch by a single blow of the hammer. Now comes the most important process of all, viz. the hardening and tempering of the blades. Upon the effectual performance of these operations depends the practical value of the articles. The Sheffield workmen have justly and deservedly acquired the very highest reputation for peculiar skill in this most difficult department of the cutlery business. The hardening of the blade is effected by heating it to bright redness, then plunging it perpendicularly into cold water, which operation renders it extremely hard, but at the same time very brittle, which is an inconvenience, of course, requiring to be remedied. This is done by the process of tempering. To this end, the hardened blades are first rubbed with finely powdered sand, to remove scales, &c. from the surface; they are then placed on an oblong tray made of steel, and on this exposed to the fire until they acquire a bright blue tint. The workman judges of the proper degree of tempering entirely by the colour, and the utmost attention is bestowed upon this point to ensure the most perfect unanimity in this respect. The hardened and tempered blades are then submitted to the manager's inspection, who applies various tests to them, and rejects any that may turn out imperfect in any one point.

The blades that have been examined and passed by the manager are next taken to the grinding mill, or, as it is technically termed, the *wheel*. Each separate shop in the building in which the grinders work is called a *hull*. The grinding is done on stones of various qualities and sizes, according to the kind of articles to be ground. The rough grit stones come mostly from Wickersley, near Rotherham; the finer and smoother grained stones, and the so-called *whitning* stones, come mostly from the more immediate neighbourhood of Sheffield. The blades of table-knives are ground on wet stones, the grinding stone being suspended, for that purpose, in an iron trough filled with water to a sufficient height to make the surface of the fluid just touch the face of the stone. The grinding stones, as well as the glazers and polishers, are turned by machinery worked by steam power. A *flat stick* is used by the grinder to keep the blade pressed to the surface of the stone. The ground blades are then glazed, which simply means that a higher degree of

Grindstones.

Chisel.

Buskin.

Flat File or Rasp.

Gauge.

lustre and smoothness is given them by grinding on a tool termed a *glazer*. This consists of a wheel made of a number of pieces of wood, put together in such a manner that the edge or face always presents the end way of the wood, which is done to preserve the circular shape by preventing contraction of the parts. The grinding face of the wheel is covered with so-called emery cake, which consists of a composition of beeswax, tallow, and emery. The glazing wheels have a diameter of four feet. The tang of the blade is stuck into a temporary handle to facilitate the operation.

The last process to which the blades of table-knives are subjected in the grinding mill is that of polishing; this is done on circular pieces of wood covered with buff leather, with a coat of finer emery (flour emery) composition upon it, which are made to revolve with much less velocity than the grinding stones and the glazers. The ground blades are again taken to the manager, who applies several very severe tests to them, to try their temper and edge.

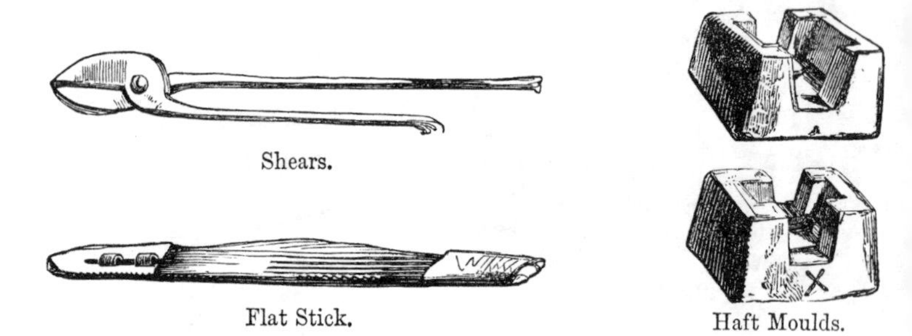

Shears.

Flat Stick.

Haft Moulds.

Knife-handles are made of horn, ivory, ebony, silver, German silver, mother of pearl, &c. Two sorts of ivory are principally used, the Egyptian and the African; the latter is the more beautiful and transparent of the two, the Egyptian looking more like horn. The tusks are sawn in appropriate lengths, which are then cut by a small circular saw into handles of the required size. The handles are properly filed, and occasionally also carved or fluted in different patterns. A variety of files are used for these purposes, such as flat files, threading files, hollow files, half round files, &c. The handle is then bored to receive the tang. The bolster of the blade having been properly filed, the tang is inserted into the bore, and fixed in by cement in the usual way. It is afterwards further secured by a German silver pin passing through the handle and tang.

The silver and German silver handles are stamped in dies. The mother of pearl handles are carved or fluted in different patterns.

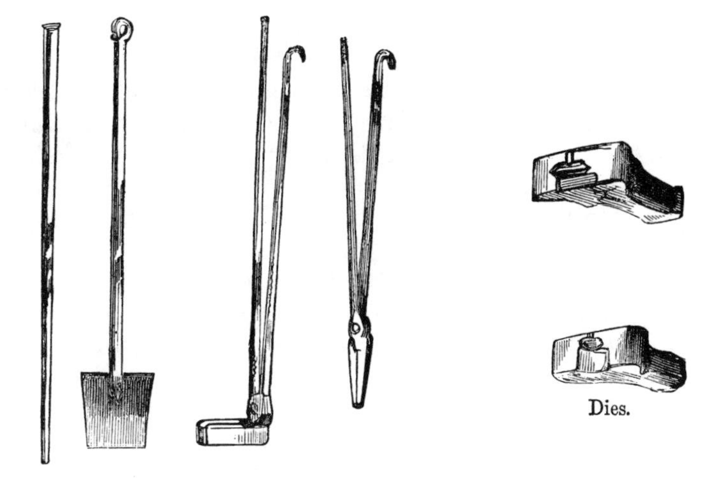

Dies.

Fire-irons. Crooked Tongs. Tongs.

The knives thus finished by the hafter are now taken once more to the manager, to undergo a final examination preparatory to their removal to the warehouse.

The forging of razors is performed by a foreman and striker in the same manner as in making the blades of table-knives. The bars or rods as they come from the tilt and rolling mill are about half an inch broad, and no thicker than is sufficient for the back of the razor. The anvil on which the razor-blades are forged is rounded at the sides; by dexterously working the blade on the rounded edge of the anvil, a concave surface is given to the sides, and the edge part thus made thinner, which saves the grinder a deal of labour. The blade having been cut off the bar, the tang is formed by drawing out the steel. The blade is then properly hardened and tempered. The last and most important process which the razor-blade has to undergo is that of grinding.

The difference in the prices of blades, make all of them of the same material, is owing entirely to the circumstance that stones of much smaller diameter are used for grinding the higher priced blades, and much more time and labour are given to the operation than is the case with the cheaper sorts.

In making a fork, the end of a steel bar is first made red-hot; it is hammered so as to give a rough approximation to the shape of the shank or tang; it is again heated, and a blow from a die or stamp gives the proper contour; the prongs are cut out by a powerful blow from a stamp of peculiar form, and the fork is finally annealed, hardened, ground, and polished. It is this process of fork grinding which has so often been made a subject for comment; the fork is ground *dry* upon a stone wheel, and the particles of steel and grit are constantly entering the lungs of the workmen, thereby ruining the health and shortening the duration of life.

Many contrivances have been devised for obviating this evil, but the fork-grinders have not seconded these efforts so zealously as might have been expected.

In making pen and pocket-knives, a slender rod of steel is heated at the end, hammered to the form of a blade, and carried through many subsequent processes. But the putting together of these hinged knives requires more time than the making of the blades, and affords a curious example of minute detail. When the pieces of bone, ivory, pearl, tortoise-shell, horn, or other substances, which are to form the outer surface of the handle, are roughly cut to shape; when the blade has been forged and ground, and when the steel for the spring is procured, the whole are placed in the hand of a workman, who proceeds to build up a clasp-knife from the little fragments placed at his disposal. So many are the details to be attended to, that a common two-bladed knife has to pass through his hands seventy or eighty times before it is finished.

A file, as every one knows, is a steel instrument, having flat or curved surfaces so notched or serrated as to produce a series of fine teeth or cutting edges, which are employed for the abrasion of metal, ivory, wood, &c.

Steel for making files being required to be of unusual hardness, is more highly converted than for other purposes, and is sometimes said to be *double converted*. Small files are mostly made of cast steel. The very large files called *smiths' rubbers* are generally forged immediately from the converted bars. Smaller files are forged from bars which are wrought to the required form and size by the action of tilt-hammers, either from blistered bars or from ingots of cast steel. These bars are cut into pieces suitable for making one file each, which are heated in a forge-fire, and then wrought to the required shape on an anvil by two men, one of whom superintends the work while the other acts as general assistant.

The next operation upon the blanks which are to be converted into files is that of *softening* or *lightening*, to render the steel capable of being cut with the toothing instruments. This is effected by a gradual heating and a

File Cutting.

File Cutting.

gradual cooling. The surface is then rendered smooth, either by filing or grinding.

The cutting of the teeth is usually performed by workmen sitting astride upon a board or saddle-shaped seat in front of a bench, upon which is fixed a kind of small anvil. Laying the blank file across the anvil, the Cutler secures it from moving by a strap which passes over each end and under his feet, like the stirrup of the shoemaker. He then takes in his left hand a very carefully ground chisel made of the best steel, and in his right a peculiarly shaped hammer. If the file be flat, or have one or more flat surfaces, the operator places the steel chisel upon it at a particular angle or inclination, and with one blow of the hammer cuts an indentation or furrow completely across its face from side to side, and then moves the chisel to the requisite positions for making similar and parallel cuts. If it be a half round file, as a staight-edged chisel is used, a number of small cuts are necessary to extend across the file from edge to edge. So minute are these cuts in some kinds of files, that in one specimen about ten inches long, flat on one side and round on the other, there are more than 20,000 cuts, each made with a separate blow from the hammer, and the cutting tool being shifted after each blow. The range of manufactures afford few more striking examples of the peculiar manual skill acquired by long practice.

Several highly ingenious machines have been contrived for superseding the tedious operation of file cutting by hand; but suited as the process may appear to be for the use of machinery, it has been found to present such great difficulties, that we believe no file-cutting engine has been brought successfully or extensively into operation. One very serious difficulty arises from the fact that, if one part of the file be either a little softer than the adjacent parts, or a little narrower, so as to present less resistance to the blow of the hammer, a machine would, owing to the perfect uniformity of its stroke, make a deeper cut there than elsewhere.

After the files have been cut, the steel is brought to a state of great hardness; this is effected in various ways, according to the purpose to which the file is to be applied; they are generally coated with a sort of temporary varnish, then heated in a stove, and then suddenly quenched. After hardening, the files are scoured, washed, dried, and tested.

It will be seen that the tools employed by the Cutler are few, and consist mostly of the hammers, moulds, dies, anvils, grinding stones, and others already mentioned.

COTTON MANUFACTURER.

CARDING.

THE extremely valuable substance, called Cotton, which is now raised in such abundance as to furnish the cheapest and most extensively-used clothing, is produced in the seed vessels of the cotton plant, of which there are many varieties; some are herbaceous annual plants, growing from eighteen to twenty-four inches high; others, shrubs about the size of our currant bushes, and of from two to ten years' duration; whilst a third kind attain the growth of small trees, with a height of from twelve to twenty feet.

The leaves of the cotton plant are of a bright dark green colour, deeply divided into five lobes; the flowers are large and showy, of a bright sulphur or lemon colour, and closely resemble in appearance and botanical structure those of the single hollyhock; each flower is succeeded by a triangular three-celled seed vessel, which attains the size of a small walnut, and when ripe bursts open from the swelling of the cotton contained in the three cells; the seeds, which are rather larger than those of grapes, are inclosed in the cotton wool, which adheres very firmly to them. One variety of

Cotton Plant.

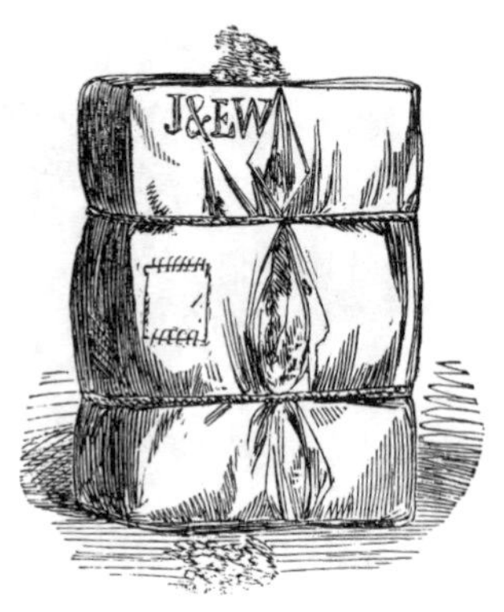

Bale of Cotton.

cotton, cultivated in China, and some parts of America, has a yellow tint; this tint it preserves when woven into the fabric called "nankeen."

The cotton plant is largely cultivated in India, China, United States, West Indies, on the shores of the Mediterranean, and, in short, in almost all the warmer parts of the world; it flourishes readily in soils too poor for the growth of grain and other crops, and succeeds perfectly well in dry seasons.

The cotton, when perfectly ripe, is gathered by women and children, the seeds and wool being picked out of the pod; it is dried in the sun, and is then ready for the removal of the seeds. In India this operation is performed by means of two parallel rollers, which are fixed in a frame at a small distance apart, so that when they are turned round the cotton is drawn through whilst the seeds, which from their size are unable to pass, are torn off and separated. With this simple machine a man can separate the seeds from about fifty pounds of cotton in a day.

In America a still more rapid process is adopted: the cotton is placed in a box, one side of which is formed of stout parallel wires, placed about one-eighth of an inch apart; by the side of this box is a roller, carrying a number of circular saws with curved teeth, which project through the wires into the box. On the roller being made to revolve, the teeth of the saws drag the cotton through the wires, the seeds remaining behind; after being thus separated, the cotton is powerfully compressed into bags, and is ready for transport to this and other manufacturing countries.

The cotton is seldom unpacked until it arrives at the mill, the purchases being all managed by samples. When it is unpacked, the first thing to be done is the sorting, and in this much care and skill are required; for the different bags furnish different qualities of cotton, and it is necessary to produce yarn of uniform quality at the cheapest rate.

In order, therefore, to equalize the different qualities, the contents of all the bags are mixed together in the following manner. A space being cleared and marked out on the floor, the cotton contained in the first bag is scattered over this space, so as exactly to cover it; the contents of the second bag are in like manner spread over the first, and the cotton in all the other bags is disposed in a similar manner; men and boys tread down the heap, which is called a *bing* or *bunker*, until at length it rises up in shape and dimensions very much like a haystack. Whenever a supply of cotton is taken from the bing it is torn down with a rake from top to bottom, by which means it is evident the contents of the different bags are collected together in a mass of uniform quality and colour. In mixing different qualities of cotton it is usual to bring together such only as have a similar length of staple. A portion of the waste cotton of the mill is also mixed in the bing, for making the lower qualities of yarn. For higher numbers, as well as for warps, a finer quality of cotton must be selected; and thus it will be seen that the formation of a bing is an important operation, the quality of the goods produced depending upon it.

In this state the cotton contains sand, dirt, and other impurities, and the fibres are matted together by the pressure they were subjected to in packing. To open the fibres and get rid of the sand, &c. the cotton is put into a machine called a *willow*. This consists of a box or case, containing a conical wooden beam, studded over with iron spikes; this beam is made to turn round five or six hundred times a minute. The cotton, as it is torn down from the bing, is put in at one end of the machine, where it is caught by the spikes, tossed about with great violence, and gradually driven forward to the other end. The sand and other impurities fall out of the machine through an open grating at the bottom; the dust and lighter matter pass off through a series of wire openings, and the cleaned cotton is sent down a shoot into the room below.

If the cotton is of fine quality it is beaten, or *batted*, with hazel or holly twigs. For this purpose, it is spread on a frame, the upper part of which is made of cords and is quite elastic. A woman, with a rod about three or four feet long in each hand, beats the cotton with great violence, and so entirely separates the fibre. Any loose impurities

Opening the Cotton.

which remain fall out between the cords; seeds and fragments of seed-pods, which adhere to the cotton somewhat firmly, are picked out by hand. By this method the cotton is thoroughly opened, and made quite clean, without injuring the staple.

The coarser qualities are passed at once from the willow to the *scutching* or *blowing machine*, which does the work of batting, only in a more violent manner, and is therefore not adapted for fine qualities; but in coarser spinning is in general use, to prepare the cotton for the carding engine.

The cotton, which is still in a confused and tangled state, has now to be carded, upon the regularity and perfection of which process depends much of the success of spinning, and also the durability and beauty of the stuff to be woven. A cotton card is a sort of brush, containing wires instead of bristles. The cards are made of bands or fillets of leather, or are formed of alternate layers of cotton, linen, and india-rubber pierced with numerous holes, in which are fixed bent pieces of iron wire, called dents or teeth.

The fibres of the cotton are not yet sufficiently level to be twisted into yarn; and it often happens that the teeth of the card lay hold of a fibre by the middle and thus double it together, in which state it is unfit for spinning.

The cardings are therefore doubled and drawn out by a machine called a *drawing frame*, the principle of which depends upon different pairs of rollers revolving with different degrees of rapidity. If, however, the riband, as it leaves the carding-engine, were simply extended in length by drawing it out, it would be liable to tear across, or to be of a different thickness at different parts of its length. To prevent the tearing and to equalize the thickness, a number of cardings are joined together and drawn out to a length equal to the sum of the length of all the separate cardings.

The effect produced is the same as taking a piece of cotton wool between the finger and thumb and drawing it out many times, laying the drawn filaments over each other, before each drawing. If the cotton be then examined it will be found that all the fibres are parallel and of equal length. This effect is accomplished very perfectly in the drawing frame, which consists of a number of rollers arranged in what are called *heads*, each head consisting of three pairs of rollers, of which the second pair moves with greater speed than the first, and the third moves quicker than the second.

By the process of doubling and drawing, the cotton is formed into a loose porous cord, the fibres of which are

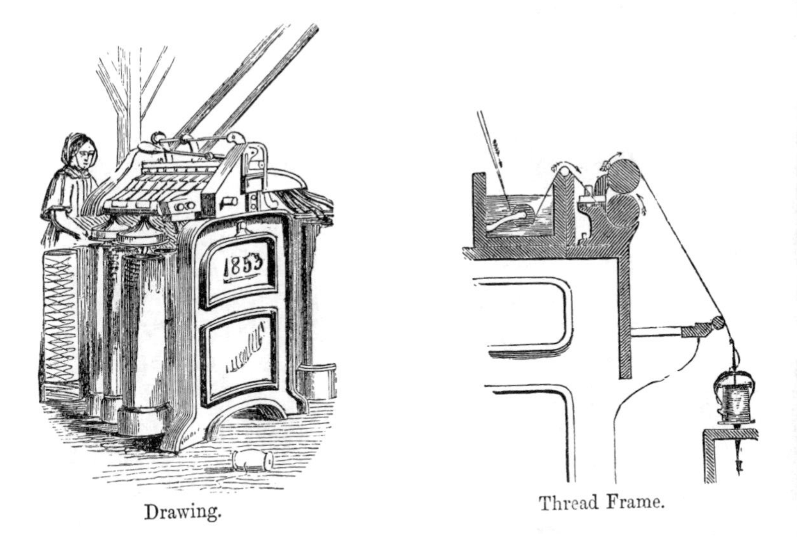

Drawing.

Thread Frame.

arranged side by side. This cord is still too thick for yarn, but it cannot be reduced in size by drawing merely, for if this were attempted it would break; a slight twist is therefore given, which by condensing the fibres allows the drawing to proceed. This is the commencement of the spinning process (which is, in fact, little more than a combination of drawing and twisting) and is called *roving*.

The bobbin-and-fly frame is an exceedingly complicated

machine, although the objects to be accomplished by it are sufficiently simple; namely, to give the roving a slight twist, and then to wind it on the bobbin. The first is easily done by the revolutions of the spindle; the second

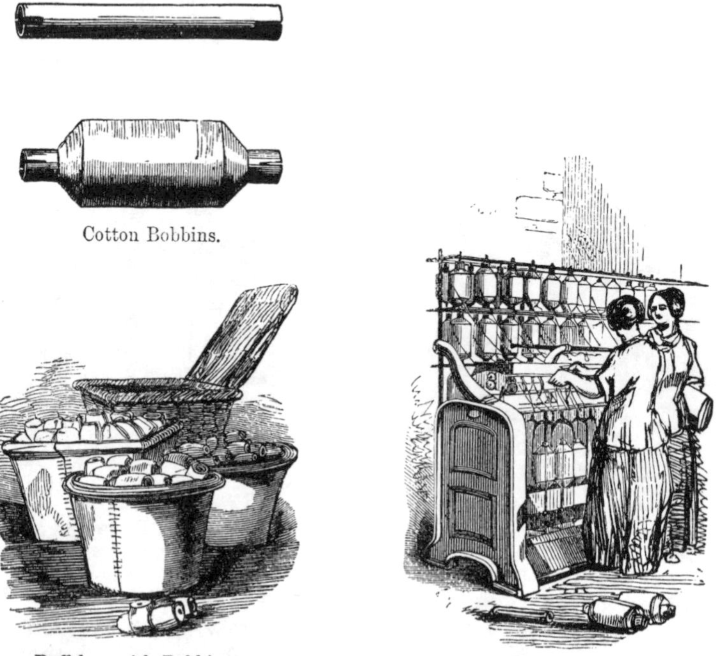

Cotton Bobbins.

Buffaloes with Bobbins.

Roving.

is more difficult. It is scarcely necessary to explain that the bobbins now under notice differ in no way from the reels in common use, except in being of very large size. The spindle which holds the bobbin is a round steel rod, driven by a small cog-wheel, fastened on the lower part of the spindle. The bobbin is slid upon the spindle, and the small bed or platform on which it rests is made to revolve by another series of small wheels. The spindle has two arms, called the *fly* or *flyer*. This fly is fixed on the top of the spindle in such a way that it can be taken off in an instant, for the purpose of putting on or taking off the

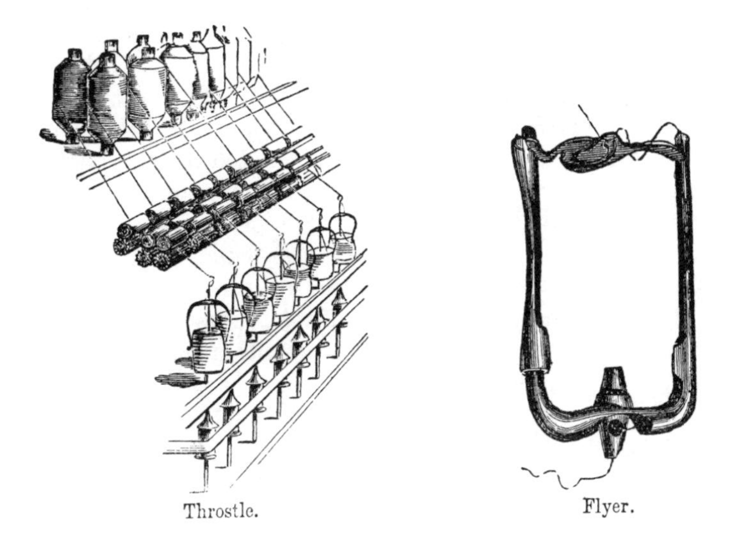

Throstle. Flyer.

bobbin. One arm of the fly is hollow, the other solid, and all this serves to balance the machinery. One machine contains from thirty to a hundred and twenty spindles, which, for economy of space, are placed in two rows, each spindle in the back row standing opposite the space left between two spindles of the front row. The action of the machine is this:—The sliver having been drawn by the rollers, is twisted by the rapid revolutions of the spindle into a soft cord or roving; this enters a hole in the top of the spindle, and passes down the hollow arm of the fly; it is then twisted round a steel finger, which winds it on the bobbin with a certain pressure.

The *throstle machine* is usually made double, a row of bobbins, spindles, &c. occupying each side of the frame. The bobbins filled with rovings from the bobbin-and-fly frame, are mounted at the upper part of the frame in two ranges. The roving from each bobbin passes through three pairs of drawing rollers, where it is stretched out to the requisite fineness. On quitting the last pair of rollers, each thread is guided by a little ring or a notch of smooth glass, let into the frame, towards the spindles, which revolve with great rapidity, producing by the motion of their flyers through the air a low musical hum, which is supposed to have given the name of throstle to this machine. The roving, which may now be called yarn, passing through an eyelet formed at the end of one of the arms of the flyer, proceeds at once to other bobbins.

The yarn is wound upon the bobbins by a curious contrivance. The bobbin fits very loosely upon the spindle, and rests on its end upon a kind of platform. The bobbin is not connected with the spindle, except by the thread of yarn, which has to be wound; therefore, as soon as the flyer is set spinning, the thread drags the bobbin after it, and makes it follow the motion of the spindle and fly; but the weight of the bobbin, and its friction on the platform, which is promoted by covering the end with coarse cloth, causes it to hang back, and thus the double purpose is served, of keeping the thread stretched and winding it on the bobbin much more slowly than the flyer revolves. The yarn is equally distributed on the bobbin by a slow up-and-down movement of the platform.

These effects are the same as were produced by the bobbin and fly-frame, but in the throstle they are attained by simpler means. In the former machine a distinct movement caused the bobbin to revolve quicker than the spindle. In the throstle the bobbin is made to revolve by the pull of the yarn, which is now sufficiently strong for the purpose; but the roving in the bobbin-and-fly frame would not bear

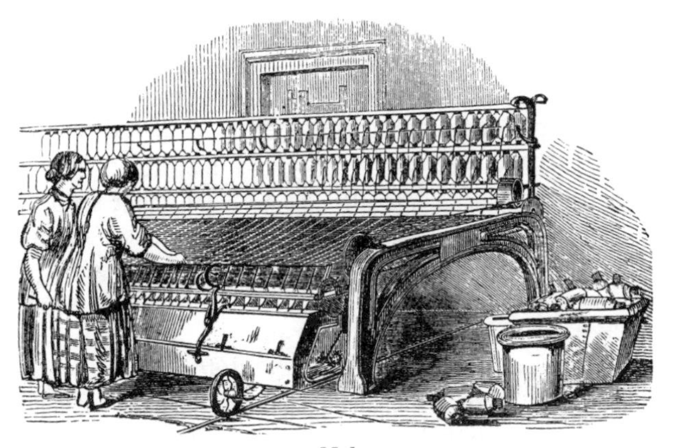

Mule.

the strain. The throstle is not often employed for very fine spinning, because fine yarn would not bear the drag of the bobbin; but in mule-spinning the yarn is wound at once upon the spindles without any strain. In the *mule* the rolling is first drawn by the usual system of rollers, and then stretched by a moveable carriage. The effect of first drawing and then stretching is to make the yarn finer and more uniform, as will be explained presently. The spinning mule is the most interesting and impressive spectacle in a large cotton mill—on account of its vast extent, the great quantity of work performed by it, and the wonderful complication and ingenuity of its parts.

The spinning-mule consists of two principal portions: the first, which is fixed, contains the bobbins of rovings and the drawing rollers; the second is a sort of carriage, moving upon an iron railroad, and capable of being drawn out to a distance of about five feet from the fixed frame. This carriage carries the spindles, the number of which is half that of the bobbins of rovings. Motion is given to the spindles by means of vertical drums, round which are passed slender cords, communicating with the spindles. There is one drum to every twenty-four spindles.

The carriage being run up to the point from which it starts in spinning, the spindles are near to the roller-beam; the rollers now begin to turn, and to give out yarn, which is immediately twisted by the revolution of the spindles; the carriage then moves away from the roller beam, somewhat quicker than the threads are delivered, so that they receive a certain amount of stretching, which gives value to this machine. The beneficial effect is produced in this way,—when the thread leaves the rollers, it is thicker in some parts than in others, and those thicker parts not being so much twisted as the thinner ones are softer, and yield to the stretching power of the mule, so that the twist is equalised throughout, and the yarn becomes more uniform. When the carriage has *completed* a *stretch*, or is drawn out from about fifty-four to sixty-four inches from the roller-beam, the drawing rollers cease to give out yarn, but the spindles continue to whirl until the threads are properly twisted. In spinning the finer yarns, the carriage sometimes makes what is called a *second stretch,* during which the spindles are made to revolve much more rapidly than before. The drawing, stretching, and twisting, of a length of thread being thus completed, the mule disengages itself from the parts of the machinery by which it has hitherto been driven, and the spinner then pushes the carriage with his knee back to the roller-beam, turning at the same time with his right hand a fly wheel, which gives motion to the spindles. At the same time a copping wire, as it is called, is pressed upon the threads by the spinners' left hand, and

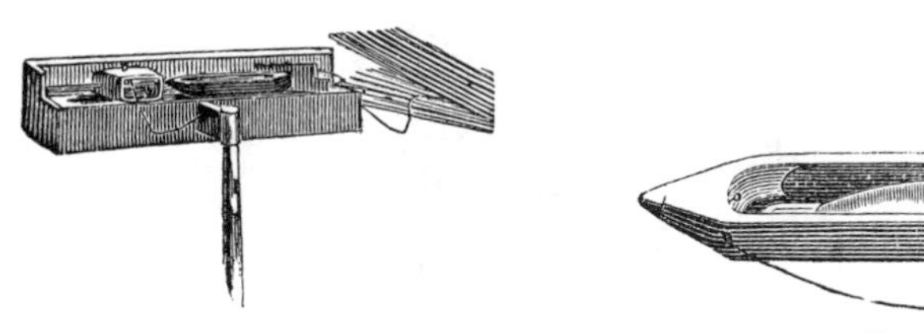

Machinery for moving Shuttle. Shuttle for Power Loom.

they are thus made to traverse the whole length of the spindle, upon which they are then wound or *built* in a conical form, which is called a *cop.* These cops are used for placing in the *shuttle* in weaving, and form the weft or short cross threads of the cloth.

The yarn is now disposed of in various ways, according to the use for which it is intended; but it is often found convenient to make it up into hanks. When the yarn is completed it is usually sent to the doubling and twisting mill, for the purpose of being converted into what is now properly called thread. Although we are accustomed to apply the word thread to a thin narrow line of any fibrous material, the manufacturer limits the term to that compound cord produced by doubling or twisting two or more single lines. The single line he calls yarn, two or more single lines, laid parallel and twisted together, he calls thread; and of this there are many varieties, such as *bobbin-net-lace-thread*, stocking-thread, sewing-thread, &c.

Gassing.

In fine spinning, the yarn, when doubled, is for some purposes *singed* or *gassed*, in order to get rid of the loose fibres, and to make it more level and compact. The process of singeing yarn strikes a stranger as being more remarkable than anything else in the mill. In a long room in the upper part of the mill, or in a shed attached to it, are several tables, lighted up with a large number of jets of flame, about twelve inches apart, producing a singular but pleasing effect. Above each flame is a little hood or chimney. On entering this room the smell of burnt cotton is immediately perceived, and on approaching the table, one is surprised to see a fine delicate thread crossing each flame in two or three directions, and apparently at rest; but on following the course of this thread, it is found to proceed from one bobbin, which is rapidly spinning round, and to pass through the flame to another bobbin, which is also in rapid motion. It is then seen that the thread is also moving at a rapid rate, by which means alone does it

Warping Machines.

escape being consumed. The thread is led over pulleys, so as to pass two or three times through the flame, which singes off the loose fibres, converting them into a reddish powder or dust, which, if blown about and inhaled, would do great injury to the lungs; this is why the gassing-room is in a remote or retired part of the building, to prevent the air being disturbed by the bustle of the heavier parts.

When cotton is intended to be woven into a fabric, such as calico, &c. the first operation consists in laying the requisite number of threads together to form the width of the cloth; this is called *warping*. Supposing there to be 1,000 threads in the width of a piece of cloth, then the yarn, wound on the bobbins as it leaves the hands of the spinner, must be so unwound and laid out as to form 1,000 lengths, constituting, when laid parallel, the warp of the intended cloth. The ancient method was to draw out the warp from the bobbins at full length on an open field (and this is still practised in India and China), but the *warping-frame* is now employed, in which the threads are arranged, by means of a frame turning on an upright centre. When the warp is arranged round this machine, the warper takes it off and winds it on a stick into a ball, preparatory to the process of beaming or winding it on the beam of the loom. The threads in this latter process are wound as evenly as possible on the beam; a separator, ravel, or comb being used to lay them parallel, and to spread them out to about the intended width of the cloth. Arrangements are then made for *drawing* or attaching the warp-threads individually to certain mechanism of the loom. In this process all the threads are attached to stays fixed to two frames, called *treadles*, in such a manner that all the alternate threads (1st, 3d, 5th,) can be drawn up or down by one heddle, and all the rest (2d, 4th, 6th, &c.) by the other.

There are three movements attending every thread of weft which the weaver throws across the warp. In the first place he presses down one of the two *treadles*, by which one of the two heddles is depressed, thereby forming a kind of opening called the *shed*. Into this shed, at the second movement, he throws the shuttle, containing the weft-thread, with sufficient force to drive it across the whole web. Then at the third movement he grasps the *batten*, which is a kind of frame, carrying at its lower edge a comb-like piece, having as many teeth as there are threads in the warp, and with this he drives up the thread of weft close to those previously thrown. One thread of weft is thus completed, and the weaver proceeds to throw another in a similar way, but in a reverse order, that is, by depres-

Weaving by hand.

sing the left treadle instead of the right, and by throwing the shuttle from left to right, instead of from right to left. In the commonest mode of weaving the shuttle is thrown by both hands alternately; but, about a century ago, John Kay invented the *fly-shuttle*, in which a string and handle are so placed that the weaver can work the shuttle both ways with one hand.

In 1678 M. de Gennes invented a rude kind of weaving-machine, intended to increase the power of the common loom; and other looms were invented, which were to be worked by a winch, by water power, or by some contrivance more expeditious than common hand-weaving; but a greater step in advance was made by the invention of Dr. Cartwright's *power-loom* in 1785. One cause which delayed the adoption of power-looms was the necessity for stopping

Power Loom.

the machine frequently, in order to dress the warp with paste or size as it unrolled from the beam, which operation required a person to be employed for each loom, so that there was no saving of expense. But the successive inventions of Radcliffe, Horrocks, Marsland, Roberts, and others, have since brought the dressing-machine and the power-loom to a high state of efficiency.

Taking a piece of calico as the representative of plain

fabrics generally, the mode of proceeding in power-loom factories may be shortly sketched as follows. The warping-frame is so arranged as to be worked by steam-power, and to bring the yarns into a parallel layer, which is transferred to the dressing-machine. This latter is a large piece of mechanism, in which the threads dip into paste on their way to the warp-beam: undergoing a process of brushing after the dipping. After this dressing the drawing and mounting for the loom are attended to. When the warp is properly arranged in the loom, steam-power does all the rest; it forms the shed or division of the warp into two parts, it throws the shuttle, it drives up the weft with the batten, it unwinds the warp from the warp-roller, and winds the woven material on the cloth-roller.

Spinning by hand.

THE TAILOR.

WORKSHOP

It would perhaps be too much to say that the more civilized a nation becomes, the greater is the attention bestowed upon dress; since it has happened that in countries not very far removed from barbarism, vast importance has been given to external display, and robes and trappings have been used to cover the savagery which had only just learned to delight in the pomp and magnificence of costly ornaments. It is certain, however, that in all civilized nations dress is more than mere clothing, and has a significance beyond the mere utility of protecting the body from cold or heat, and adding to our physical comfort. It is an old saying that some people may be clothed, but that they are never dressed; and the meaning of this is, that dress is frequently an expression of character, and will even make known the disposition of the wearer. In the same way a change of *fashion* is often an indication of an alteration in the manners and way of living of a whole nation.

It is the business of the Tailor, then, not only to make garments, but to study the prevailing fashion, and, indeed, to advise what alterations or slight differences in the cut and colour of clothes will be best suited to different people, since on the way in which our clothes are made our personal appearance will very greatly depend, and personal appearance is of no little importance, since there are few people who are not strongly influenced by it.

The trade of the Tailor is one of which very little can be said in the way of explanation, since it mainly consists in cutting out cloth to the shapes necessary to be applied to each other in order to make the various garments; and as this cannot be described without numerous *diagrams*, and even with them could not easily be understood, we must be satisfied to quote the instructions of a practical Tailor on the subject of sewing.*

And, first as to the different sorts of stitches, which are:—the basting-stitch, the back and fore-stitch, the back-stitch, the side-stitch, and the fore-stitch; also the back

* From the Industrial Library.

pricking-stitch, the fore pricking-stitch, the serging-stitch, the cross-stitch, and the button-hole-stitch; besides which there is a distinct kind of stitch for hemming, filling, stotting, rantering, fine-drawing, prick-drawing, over-casting, and also for making what are called covered buttons.

The basting-stitch is a long and slight stitch, intended to be merely temporary, or to fasten together some of the inner and concealed parts of the garment. It is commonly used to keep the work in its proper position while being sewed.

The back and fore-stitch is made, as the name implies, by the union of back stitching and fore stitching; in this stitch the needle is first put through the cloth, and turned

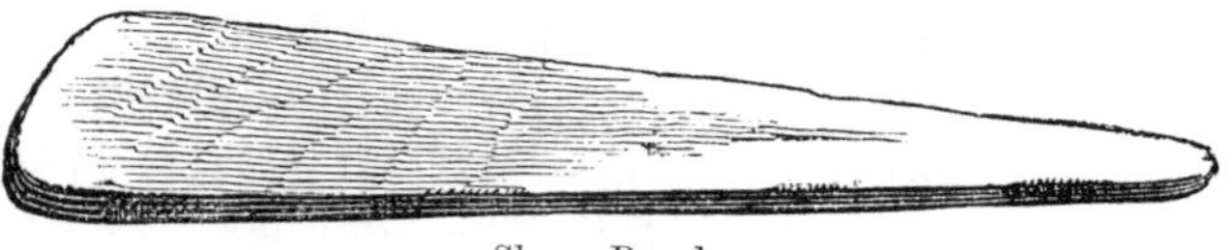

Sleeve Board.

up in as short a space as is possible, so as to make a neat and strong stitch when completed; it is then put through the cloth again in the same place as at first, and again turned up, taking care that it passes through the cloth as nearly as possible within the same space as before. This being done, the first back-stitch is completed. The second stitch is made by passing the needle forward upon the surface of the cloth, but without taking hold of it, over a space equal to the length of the first stitch; the needle is again put through the cloth, turned up, and brought back to the place where it was last put through, so as to form another back-stitch; which is followed by another putting of the needle forward, or, in plainer terms, another fore-stitch, and so on in the same order, until the seam is finished. This kind of stitch is used for sewing linings, pockets, flannel garments, and other thin fabrics. There is no need to say much respecting the back-stitch, as this may be understood from what is said above respecting the first stitch in back and fore-stitching. This stitch is used

Goose. French Chalk. Thimble. Measure Book. Flat Iron. Rule.

for seams where strength is required; it is also used for ornament instead of the side-stitch, but in this case it must be very neatly and regularly made.

The side-stitch is used for the edges of garments, to keep them from rolling over, or from being drawn out of shape. It is always intended for ornament as well as use, and requires a very quick eye and a careful hand to do it well. In this stitch the needle is put through the cloth a little above or below the place from which it came out in the former stitch, but it must be at a very little distance from this place, or the sewing-silk will be visible on the surface of the cloth, which is a great blemish, and yet it must be far enough away from where it came out to prevent its breaking through, in which case the stitch is lost both as to use and ornament. Care must also be taken that the stitches are at regular distances from each other, and that the whole of them are placed at the same distance from the edge of the cloth. In the fore-stitch, as has been already hinted, the needle, when drawn out from the seam, is always put forward, so that an equal quantity of thread, or a stitch of the same length, is visible on each side of the cloth.

Serge-stitching is done by passing the needle through the cloth from the under to the upper piece, throwing the thread over the edges of the cloth, so as to keep them closely together. It is also used to join selvages together, as also to prevent taking up more space for seams than can be spared, when the pieces are barely large enough for the required purpose. It is not, however, much used by tailors, except when no great degree of strength is required.

The cross-stitch is formed by two parallel rows of stitches, so placed as that the stitch in the upper row is opposite to the vacant space in the lower one, the thread passing from one stitch to the other in diagonal lines. It is used for keeping open the seams of such garments as require washing, and also for securing the edges from ravelling out in such fabrics as are too loosely made to allow of their edges being fastened down by the filling-stitch.

In the button-hole-stitch the needle is first put through the cloth from the inner to the outer surface, and before it is drawn out the twist is passed round the point of the needle, and kept in that position till the needle be drawn out to the full length of the twist; this forms a kind of loop, called by Tailors the "purl." at the top or edge of the opening, and when regularly made is both ornamental and useful. To increase the strength of this stitch, and also to aid in making it true or exact, a "bar" is formed on each side of the opening before the hole is begun to be worked.

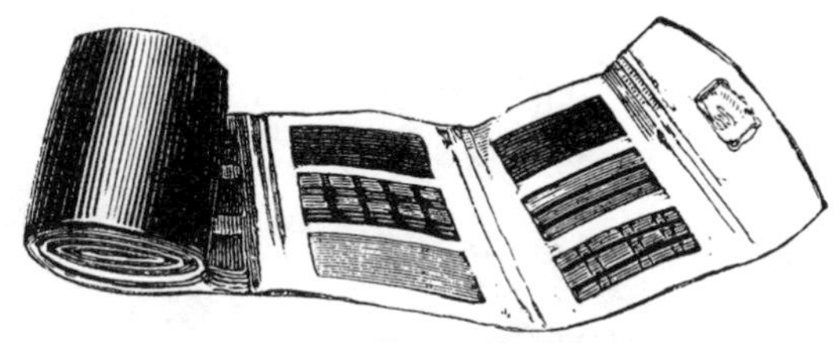

Trousers Pattern Book.

This "bar," as it is called, is made by passing the needle from one end of the opening to the other (twice or three times), so that there is a layer, if it may be so called, of twist stretching along its whole length (and on each side) upon which the whole is worked, the workman taking care to keep the "bar" as near to the edge of the opening as possible, without allowing it to come over, in which case the button-hole would be neither strong nor neat.

The filling stitch is similar to that used in hemming; the chief difference being in the direction given to the needle. In hemming, its point is directed outwards, or *from* the workman, but in filling it is directed inwards, or towards him, and in each should be a little, but only a little, slanted, in order to give the sewing a neat appearance. This stitch is used for sewing on facings, and when made with neatness, and without showing itself much on the outer side of the cloth, is considered to be ornamental, as well as useful.

Stotting (pronounced stoating) is the stitch used for joining pieces of cloth so neatly that the join shall be but little visible, and yet so strongly as to prevent the pieces from being easily parted. In this kind of seam the pieces of cloth are not laid the one upon the other, as in

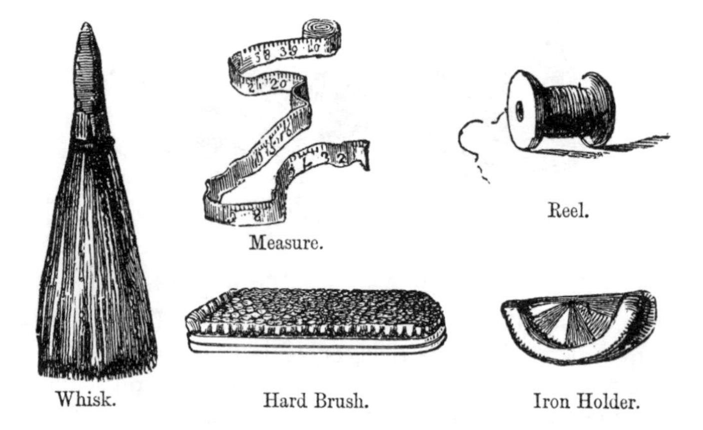

Measure. Reel.

Whisk. Hard Brush. Iron Holder.

back-stitching, but are placed side by side, the edges being carefully fitted, so as to prevent any irregularity or roughness in the work. They are then sewn together by passing the needle half through the thickness of the cloth. Care must be taken to keep the stitches as near to each edge of the cloth as can be done without incurring the danger of its breaking through. The needle is put in on the nearest edge of the two, and must not be slanted in the direction given to it, but put as straight forward as possible. The stitch should be drawn close enough home to prevent the silk thread from showing itself on the right side of the cloth, but yet not so close as to draw the edges into a ridge. If the join be as neatly made as it may be, it will, when properly pressed, be barely perceptible. This stitch is used for joining the pieces of cloth of which facings, collar-linings, and other fillings-up of the inner sides of

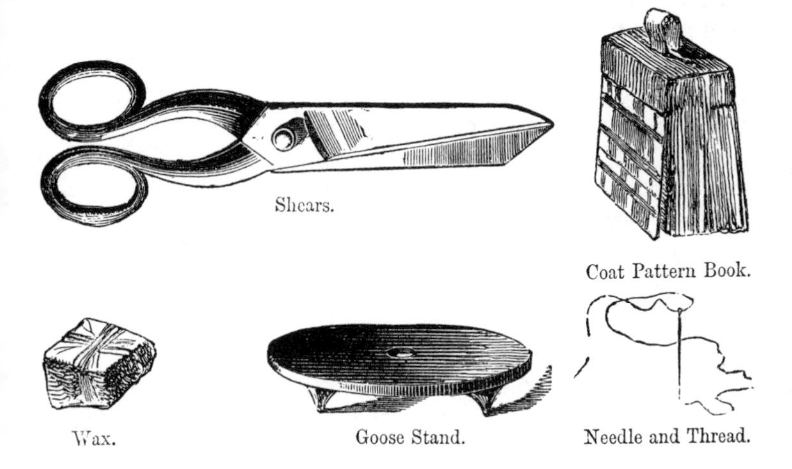

Shears. Coat Pattern Book. Wax. Goose Stand. Needle and Thread.

garments, are made, and also in other cases to prevent the taking up too much of the cloth by making a back-stitched seam.

Rantering, like stotting, is intended to conceal a join in the cloth. Here, however, it is requisite to make a strong as well as a neat joining; and therefore a seam is first sewn with a fore-stitch, and then the rantering-stitch is worked upon or over this seam. It should be worked with a very fine silk thread, or with twist that has had one of the strands taken out. The needle should be both long and slender, and must be passed forwards and backwards over the seam, so as to catch hold of its two sides, and draw them closely together. But in doing this care must be taken not to take a deep hold of the cloth : the nap or wool is all that should be taken hold of, and this must be done with a light hand, while the stiches must be placed close to each other, so that the seam may be well covered with wool ; when this is done, the seam has to be "rubbed up," that is to say, it must be held between the fore-finger and thumb of each hand, these being placed upon the fore-stitching, and its two edges brought as closely together as possible. The rantering must then be slightly carded or scratched backwards and forwards with the point of a needle, in order to bring the wool out again where it has been drawn in with the stitch ; the seam is then ready for pressing, and, if this operation be properly performed, will be as much concealed as may be necessary ; while it will be much stronger than if it had been merely back-stitched.

In fine-drawing, the stitch is formed in the same manner as in rantering, but there is a difference in the way of placing the pieces that are to be joined, *i.e.* if they be separate pieces, for this stitch is mostly used to close up places that have been accidentally cut, or torn ; the two edges of the place requiring to be fine-drawn are first trimmed by cutting away the loose threads or ends of the cloth which may be upon them ; they are then placed and kept in as level or flat a position as is possible, either with the fingers, or by fastening them to a piece of stiff paper. The needle should be both very small and long, and the thread used, whether it be of silk or twist, should be very slender. Greater care is here necessary than in rantering, to avoid taking a deep hold of the cloth; the needle should be passed forwards and backwards, over the opening, and the thread should be drawn no closer or tighter than is quite needful in order to hide it in the wool. The stitches must be placed as near to each other as is possible, so as to prevent the edges of the cloth from being visible between them; if it be needful to make a strong as well as a neat joining, the fine-drawing should be repeated on the under side of the cloth, but here it will not be needful to put the stitches so close together. When the fine-drawing is done it must be pressed, but with as light a hand and in as short a time as is practicable, otherwise the sewing, however neatly done, will be visible, and so far as it is so, the design of the fine-drawing stitch will not be answered.

The stitch called prick-drawing is now but seldom used, yet it may be proper to notice it briefly. When this stitch is intended to be employed, the edges of the cloth are first stotted together, after which the needle is passed backwards and forwards in diagonal lines, under the stotting, so as to make the join more strong and durable than it can be made by merely stotting the pieces together.

This stitch is used where the cloth is very thick, or hard and unyielding, and, consequently, where the stotting-stitch would quickly give way without this support. It is also better than a back-stitch seam for cloths of this description, inasmuch as it can be made to lie more flat, and thus to be more neat in its appearance, than a common seam.

Overcasting is used merely to secure the edges of thin and loose fabrics from "ravelling out." In using it, the edges of the cloth, whether it be woollen, linen, or cotton, are first trimmed clear of the loose threads; the needle is then passed through the cloth in a forward direction, at about the distance of one-eighth part of an inch from the edge of the cloth, and when drawn out it is carried (from the left to the right, and not, as in other stitches, from the right to the left) about a quarter of an inch; it is then again put through, and on being drawn out it is made to pass over the thread leading from the preceding stitch, so as to form a kind of loop on the edge; which loop secures the edge from becoming too much frayed, or ravelled.

All the tools that the apprentice or even the journeyman requires may be bought for a few shillings. A yard of linen for a *lap-cloth;* two pairs of *scissors,* one pair moderately large, for common use, and the other small, for button-holes; a *thimble;* a small piece of *bees-wax;* and three-pennyworth of *needles,* are all that he will have occasion to buy so long as he is not a master, or a journeyman working at home, when he must procure a *sleeve-board* and an *iron.* The more expensive part of even these few implements, viz. the scissors, will, with tolerable care, last for a number of years with only the trifling expense of being occasionally sharpened by the cutler.

All the implements used by the Tailor are so well known as to need no particular discription; the *sleeve-board* is used to place in the sleeve of a coat while the seams are pressed with the heated *iron* or the *goose.* In the *measure book* the dimensions are written when measurement of a customer is made, and the *French chalk* marks the direction in which the cloth is cut to the pattern of the various shapes, which are afterwards sewn together to make complete garments.

THE TANNER.

THE PITS.

LEATHER is a substance universally used amongst civilized and very generally amongst barbarous nations; it is made from the skins of animals, which are tanned, or prepared with some substance, having the power of converting the perishable skin, that decays readily when wet or moist, into a lasting and comparatively imperishable leather.

The preparation of skins by tanning or other similar processes has been practised from the earliest times; and although it has engaged the attention of several scientific men, and has been the subject of many curious experiments, it has received less alteration from recent improvements in chemical science than many other manufacturing processes. Several plans, which have been suggested with a view to expediting the process, which on the old system is a very tedious one, have been found to injure the quality of the leather, and have therefore been wholly or partially abandoned; and others, which appear to be more successful, are as yet adopted by a few manufacturers only.

The larger and heavier skins operated upon by the Tanner, as those of bulls, buffaloes, oxen, and cows, are technically distinguished as *hides*, while the name *skins* is applied to those of smaller animals, as calves, sheep, and goats. The process necessary to convert hides into the thick hard leather used for the soles of boots and shoes, and for similar purposes will first be noticed. The hides are brought to the Tanner either in a fresh state, when from animals recently slaughtered, or, when imported from other countries, dried or salted, and sometimes both, for the sake of preserving them from decomposition. In the former case the horns are removed, and the hide is scraped to cleanse it from any small portions of flesh or fatty matter that may adhere to the inner skin; but in the latter it is necessary to soften the hides, and bring them as nearly as possible to the fresh state, by steeping them in water, and repeated rubbing or beating. After this the hair is removed, sometimes by steeping the hides for several days in a solution of lime and water, which has the effect of loosening the hair and epidermis, or outer skin; and sometimes by suspending them in a close chamber called a smoke-house, heated a little above the ordinary temperature of the atmosphere by means of a smouldering fire, in which case the epidermis is loosened by a very slight putrefaction. In either case, when the hair and epidermis, or cuticle, are sufficiently loosened, they are removed by scraping with a curved knife, the hide being laid upon a convex bench or *beam*.

The hides are prepared for the actual tanning, or immersion in a solution of bark, by steeping them for a few days in a pit containing a sour solution of rye or barley flour, or in a very weak menstruum, consisting of one part of sul-

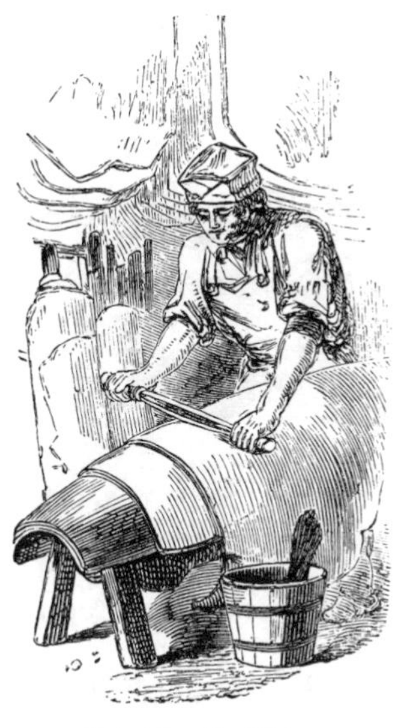

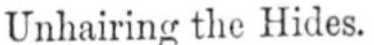

Unhairing the Hides.

Striking the Hides.

phuric acid mixed with from five hundred to a thousand parts of water. By this process, which is called "raising," the pores of the hides are distended and rendered more susceptible of the action of the tan.

Oak-bark is the substance most commonly used to supply the astringent principle, and it is crushed or ground to powder in a *bark-mill.* In the old method of tanning, which is not yet entirely abandoned, the hides and powdered bark were laid in alternate layers in the *tan pit*, which was then filled with water to the brim. After some months the pit was emptied, and refilled with fresh bark and water; and this process was repeated whenever the strength of the bark was exhausted. In this way, the time required for impregnating the hides varied, according to their thickness and

Bark Box. Barrow.

other circumstances, from one to four years. The process has been greatly expedited by the improvement, introduced in consequence of the experiments of M. Seguin, a French chemist, of tanning with concentrated solutions of bark, formed by passing water through a mass of powdered bark, until, by successive filtrations, it is completely deprived of its soluble tanning principle.

The variations of practice among different Tanners extend to the substance used as an astringent, as well as to the manner of applying it. Ground oak-bark, which was formerly the only material in common use, and is still the most general, produces good leather of a light fawn colour.

Valonia, of which considerable quantities are imported for the use of Tanners, produces leather of great solidity and weight, the colour of which is inclined to grey, and which is more impervious to water than that made with oak-bark. Valonia consists of the acorns of the *Quercus Ægilops*, and is brought from the Levant and the Morea. Catechu, or terra japonica, the extract of the *Acacia Catechu*, produces leather of a dark reddish fawn colour, which is light, spongy, and very pervious to water.

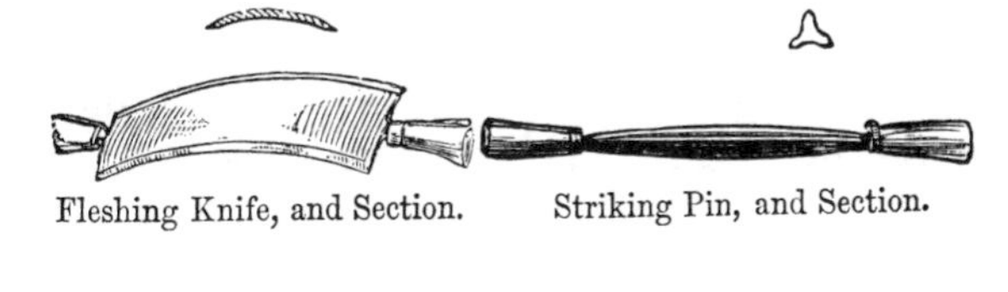

Fleshing Knife, and Section. Striking Pin, and Section.

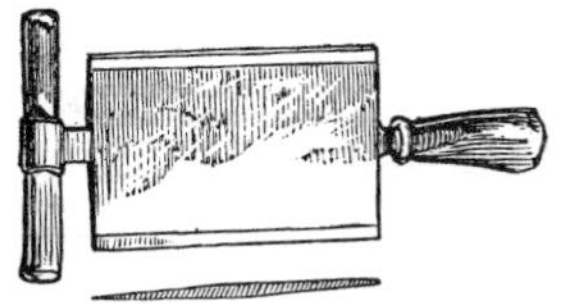

Shaving Knife, and Section.

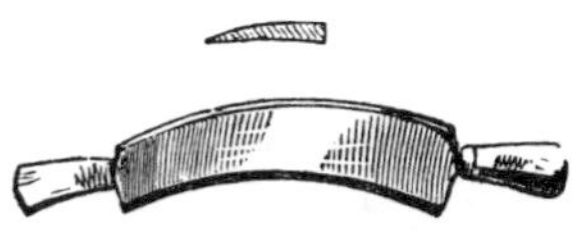

Unhairing Knife, and Section.

When the process is complete, the hides are hung up in a shed and allowed to dry slowly; and while they are drying they are compressed by beating, or rubbing, or by passing them between rollers, to give them firmness and density. A yellow deposit is now found upon the surface of the leather, to which the name of "bloom" or "pitching" is technically given.

We have hitherto alluded chiefly to the preparation of the thick hides used for sole leather, among which several varieties may be found, each distinguished by a different technical name, by which its thickness, quality, or mode of preparation is known; but the thinnest and weakest hides, as well as the skins of calves and other animals, are also prepared for use as upper leathers, in which case it is necessary to reduce their thickness by *shaving* or *paring* them down upon the flesh or inner side, before they are subjected to the action of the tanning infusions. Such hides or skins also require, after leaving the hands of the Tanner, to be rubbed, softened, and dressed by the currier, in order to bring them to the necessary degree of flexibility and smoothness. The currier also has recourse to shaving or paring with a peculiarly formed *knife*, to bring the skin to the requisite tenuity; and it is his office to blacken the surface, which, for common shoe leather, is done on the flesh side, although for some purposes leather is blackened on the outer or grain side. Horse-hides, which are comparatively weak and thin, are sometimes dressed in the latter way, under the name of Cordovan hides, from the circumstance of such leather having been formerly made at Cordova, in Spain. Calf-skins supply the quality of leather most generally preferred for the upper part of boots and shoes.

Of the thin skins prepared for ornamental purposes, many are tanned with a substance called *sumach*, prepared from a plant of the same name. The tanning is performed by sewing up each skin into the form of a bag, with the grain or hair side outwards, and nearly filling it with a strong solution of *sumach* in water. The bag is then fully distended by blowing into it, and the aperture is tied up; after which it is thrown into a large shallow vessel filled with hot water containing a little sumach. The distended bags float in this vessel, and are occasionally moved about with a wooden instrument until the solution which they contain has thoroughly penetrated their substance. Owing to the thinness of the skins and the heat to which they are

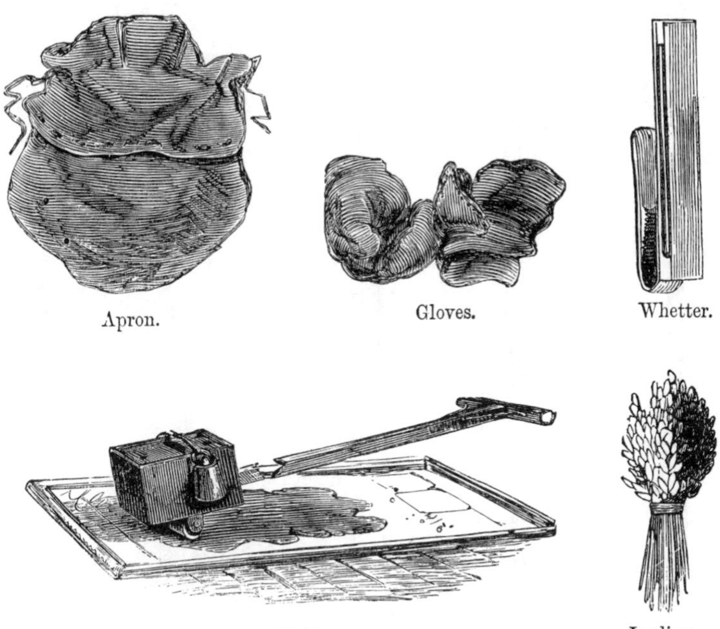

Apron. Gloves. Whetter.

Roller. Leoline.

exposed, this operation is performed in a few hours. The process is expedited by taking the bags out of the solution, and piling them upon a perforated bench or rack at the side of the tub, so that their own weight may force the confined liquid through the pores.

When the tanning is completed, the bags are opened to remove the sediment of the sumach; the skins are washed, rubbed on a board, and dried; after which they are ready for dyeing and finishing with a ridged instrument, which imparts to the surface that peculiar grain by which morocco

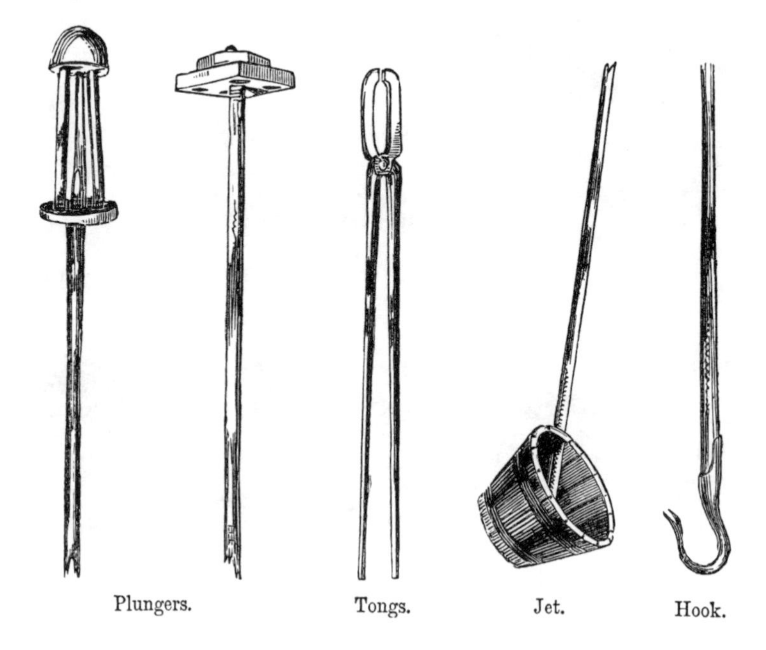

Plungers. Tongs. Jet. Hook.

leather is distinguished. An inferior kind of leather, known as "imitation morocco," is prepared in a similar manner from sheepskins.

"Tawing" is the name applied to the process by which the skins of lambs and kids are converted into soft leather by the action of alum. Of this kind of leather gloves are usually made.

The *jet* and *plungers* used for immersing the hides in the tan, the *tongs* and *hook* for removing them, and the other implements, are easily understood as applied to the various processes here mentioned.

THE SHOEMAKER.

WORKSHOP.

THE trade of the Shoemaker naturally follows that of the tanner, since leather is the material principally employed in making coverings for the feet; the tough hides of animals being used for the same purpose in countries where the art of tanning is either unknown or not practised. Scarcely any handicraft employment engages the attention of so many persons in this country as boot and shoe making. From the fashionable bootmaker to the poor cobbler, who crouches in a stall under a house in some narrow street, is a wide interval, and this interval is filled up by numerous grades. At Northampton boots and shoes are made on a very large scale for the London markets; they include chiefly the cheap varieties, but at some of the recent exhibitions of manufactures the Northampton bootmakers have exhibited specimens of workmanship which are considered to be quite equal to those of London or Paris. At Edenbridge, in Kent, and at other places, the strong coarse "hob-nailed" shoes are made, which are so much worn by waggoners and others.

The London makers import from Paris very large quantities of boot fronts, which, when combined with other parts of English manufacture, constitute many of the "French boots" which now glisten in the windows. Notwithstanding the large number of persons employed in these avocations in England, and the abundant supply of leather, there is still a considerable import of boots and shoes from abroad, chiefly France.

In the old statutes a Shoemaker is called a cordwainer, apparently a corruption of the French *cordonnier*, which means a worker of Cordova leather. The companies of Shoemakers in our ancient towns were incorporated under this name; and where some of these companies now exist, they are known by the same name. As a legal term, cordwainer is still used.

The trade, as now followed in London and other principal places, is subdivided into about twenty branches. The following may be set down as the chief: the shoeman, or maker of the sole part of the shoe; the bootman, or maker of the sole part of the boot; and the boot closer, or joiner together of the leg, vamp, &c. The labour of these is especially directed to what is called the men's line; whilst others make the ladies' shoes or boots. There are many women, too, who get a livelihood by closing the shoe, while others again follow the various sorts of binding.

The mechanical processes, after marking and cutting out the leather, consist chiefly in various kinds of strong needle-work, such as the lasting or tacking of the upper leather to the in-sole, the sewing in of the welt, the stitching to this welt of the out or top sole, the building and sewing down of the heel, and the sewing or closing of boot legs. The boot closer is the most skilful of the persons employed, and receives the highest wages.

The materials with which the Shoemaker works are generally called the *grindery*,—they are so called at least through England and Scotland, though in Dublin it is called *finding*.

"The cause of this technicality," says Mr. James Devlin, in his most interesting description of the trade of the Shoemaker in the Industrial Library, "is now, I believe, scarcely known to any one in the trade. The relation to whom I was apprenticed, a man of a very active and inquisitive turn of mind, told me its history, which it may be worth while here to relate.

"Formerly, before hemp, flax, wax, hairs, or any description of tools, were sold, as now, in shops set apart to this particular business, the shoemaker, not using the peculiar sort of stone rubber or the emery composition which he now uses to sharpen his knives upon, was in the habit occasionally of taking his knives to be ground (as the French shoemaker does at the present day) to some of the common knife-grinders of the neighbourhood. The knife-grinder having thus the Shoemaker for a regular customer, began in time to add to his usual business that of selling hemp, &c.; hence his little shop being termed the *grindery*, every thing he sold became known under this name, and is still continued."

The tools of the Shoemaker are in their collective form denominated his *kit*. Anciently, and in the old songs of the trade, they were called "St. Hugh's bones," from a now almost forgotten, though somewhat pleasant tradition. In Stow, and in Randle Holme's "Academie of Armorie," 1688, we find this term; as, also, in the still older romance of "Crispin and Crispianus;" and in two plays, "The Shoemaker is a Gentleman," and the "Shoemaker's Holiday," of the beginning of the seventeenth century.

The kit of the Shoemaker is, however, no longer now, as formerly, made up of "bones"—saint or infidel, human or brute,—but principally of good and kindly steel; purchased ready-made at the *grinder's*, or the grindery establishments before spoken of, and kept afterwards (in this country, at least, and in America) in repair and proper order by the ingenuity and care of the workman himself; though in France, and generally on the continent, much of this is done by another person, to whom such occupation is the sole means of livelihood.

Under the general term *kit* is comprehended the *pincers*, *nippers*, *hammer*, the various descriptions of *awls*, of *setting irons*, and many other articles.

It has often been a matter for great surprise that the Shoemaker should sit in such a cramped and unhealthy position at his work, crouched over his *clamps* as he holds the leather in them between his knees while engaged in

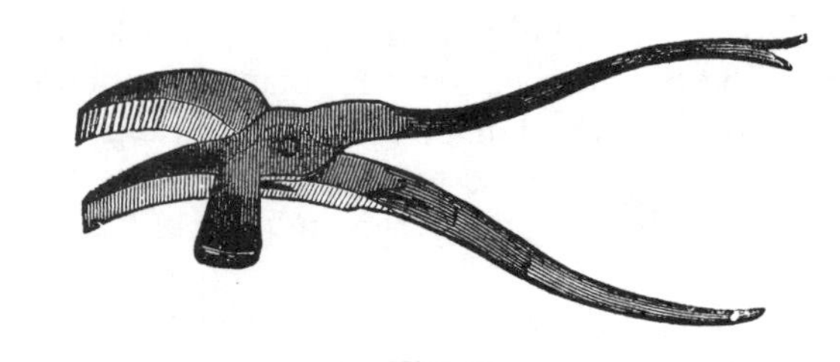

Nippers.

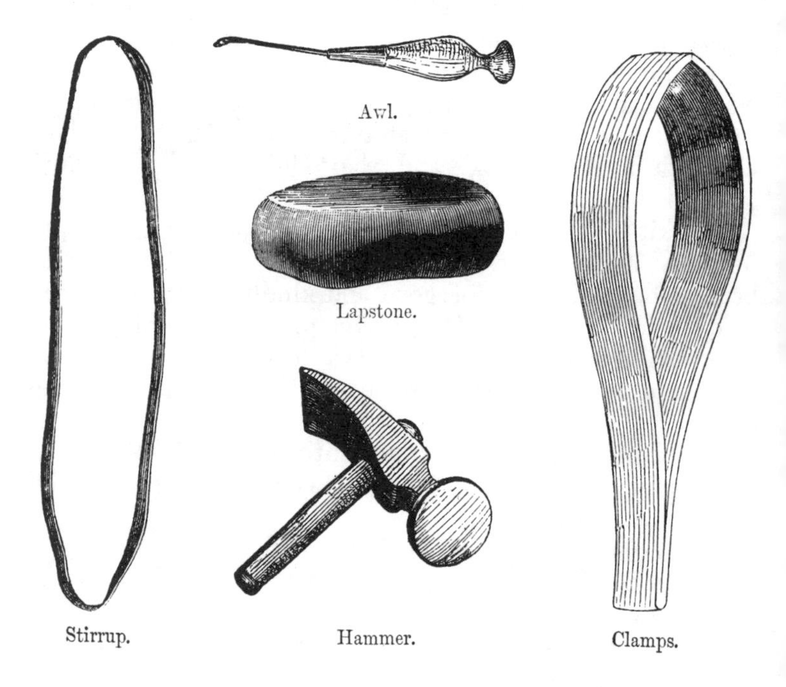

Awl.

Lapstone.

Stirrup. Hammer. Clamps.

sewing, or over the block, which he holds fast to his thigh with the *stirrup* that passes underneath his foot.

Many inventions and improvements have been made for enabling the Shoemaker to stand during a great part of his time, and some of these seem well adapted to supersede the old position, but at present they have been only partially adopted.

The Shoemaker's thread being tipped with bristles, no needle is required for sewing; but the thread itself is passed through the hole made with the *awl*. Quickly goes in the awl, and as quickly is out again, but not before the hair from the fingers of the left hand has found the passage, without being at all directed by the sight, but literally in the dark; and hence the term *blind stabbing*, the right hand hair immediately following in the opposite course, the closed thumb and fore-finger of either hand nipping at the moment the hairs from these different directions, and drawing the same as instantly out, at once completing the stitch.

A proficient closer, or closer's boy—for here, in general, the boy is even more expert than the man—will in the space of half an hour stab the four side rows and the two back rows of the counter of a boot, each inch of stitching taking about twenty stitches, and the entire work averaging about fifteen inches, three hundred stitches being thus put in in thirty minutes, or fifty every five minutes, each stitch requiring in itself six distinct operations—the skill of sight or distance, the putting in of the awl, and again its withdrawal, the putting in the left-hand hair, and again of the right, and lastly, the careful though rapid drawing, or rather twitching, out of the thread itself.

The closer needs little kit: a slip of board to *fit* or prepare the work upon; a pair of *clamps*; a *block*; a *knife*; about three *awls*, two differently-sized closing awls, and one stabbing awl; two *seam-sets*, or it may be three—one for the stabbed sides; a *stirrup;* a case of *needles* (short blunts), and a *thimble.*

The lining in all shoes, at least, but those of the very strongest kind, is entirely the work of the woman, being done with the needle, and elegantly it often is done. After

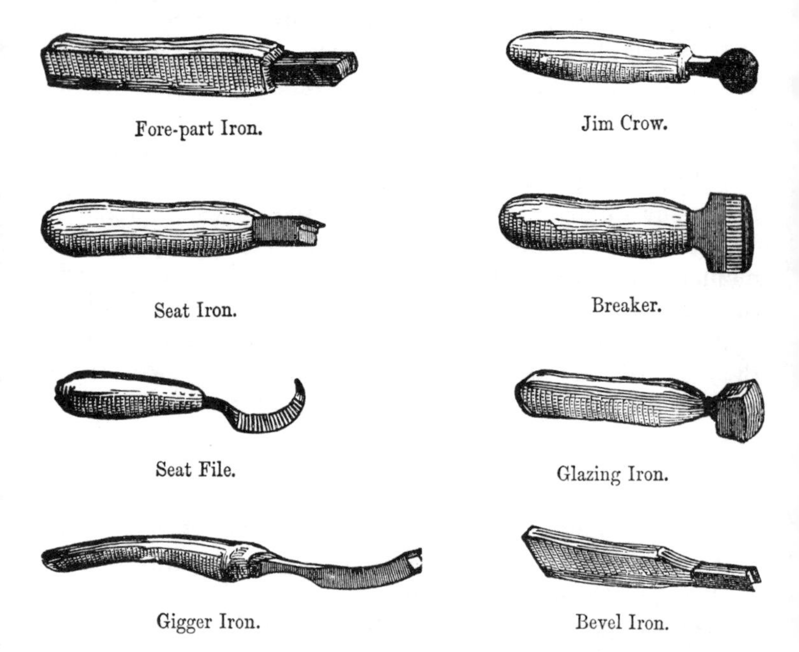

Fore-part Iron. Jim Crow.

Seat Iron. Breaker.

Seat File. Glazing Iron.

Gigger Iron. Bevel Iron.

the lining, the *upper* has to be *set;* a matter soon effected; the flat-seam-set, or, if stabbed, the stabbing-side-set, being heated at a candle (though this is not necessary, and might from the danger of the practice be well dispensed with) and a little dissolved gum being rubbed on the seam, the set is immediately to be somewhat forcibly and briskly pressed along the line of stitching, which thus takes an almost instant polish, and being also hardened, the upper becomes ready for shop; that is, to be sent to the *maker*, or shoeman, to finish by putting in the welt, soles, and other parts of the shoe; the "stitching" being effected with a square awl.

The number of tools used by the Boot and Shoemaker who combines all the branches of the trade is very great,

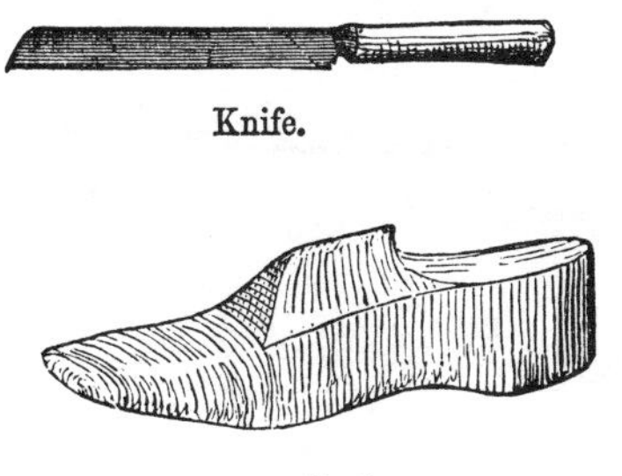

Knife.

Last

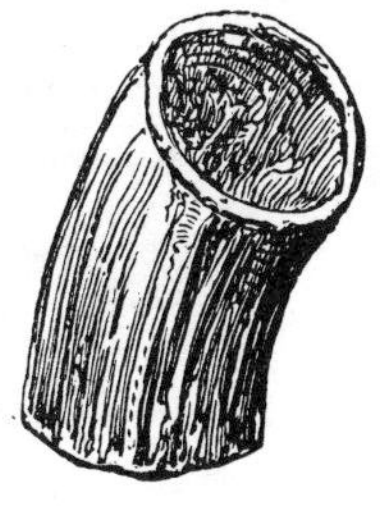

Paste Horn.

although few of them are remarkably expensive, except some of the patent metal or hardwood *lasts*, on which the shoes or boots are placed. The *lapstone* on which the leather is hammered after being damped; and the various kinds of *irons* used for rubbing, paring, or shaping the soles and other parts of the shoe, are the principal implements, beside the *awls*, *knives*, *hammers*, and *rasp*.

The *glazing iron* is used for burnishing the heel of the boot or shoe; the *paste horn* for containing the paste used in the inside lining of the shoe or boot; the *irons* for setting up the leather beside the stitching, and the *Jim-*

crow, which is a small toothed wheel running in a handle, for the same purpose. The *long stick* is used for "sleeking," or smoothing and softening the upper part of the boot or shoe, after it has been made and placed on the block, in order to take out any wrinkles that may remain.

It would be impossible to give any clear description of all the operations of the trade of the Shoemaker in the space devoted here to this particular business, and even were it attempted, no very clear idea could be conveyed of the various portions of his work; since, like the tailor, he

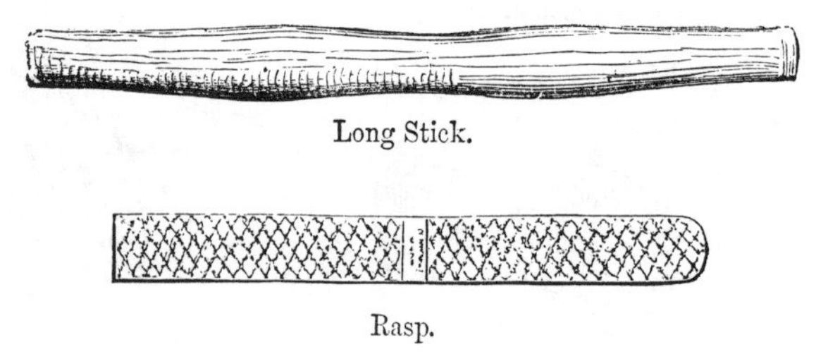

Long Stick.

Rasp.

has first to cut out the leather to the proper shape for making the various parts, which have afterwards to be put together. Indeed, beyond mentioning the uses of the various tools, most of which are simple enough, little can be said to explain the operation of the Shoemaker that would not require actual inspection of the different processes; although many of them may be understood by looking at a shoe, after having observed the accompanying pictures and having read what has been said about the tools employed.

THE SADDLER AND HARNESS MAKER.

WORKSHOP.

THE manufacture of saddles and harness for horses is one requiring very considerable skill and no little patience, since on the ability of the Saddler depends not only the health and comfort of the horse, but the safety of the rider.

The beauty and finish of the harness and its appurtenances are so essential to the proper appearance presented by the whole equipage, that a very great deal of attention is now given even to the smallest details, such as buckles, mounts, and ornamental sewing on straps and traces.

The operations in the trade of the Saddler are so similar to those of the shoemaker, as far as the stitching and cutting of leather are concerned, that they require very little description. The most difficult part of his business is the

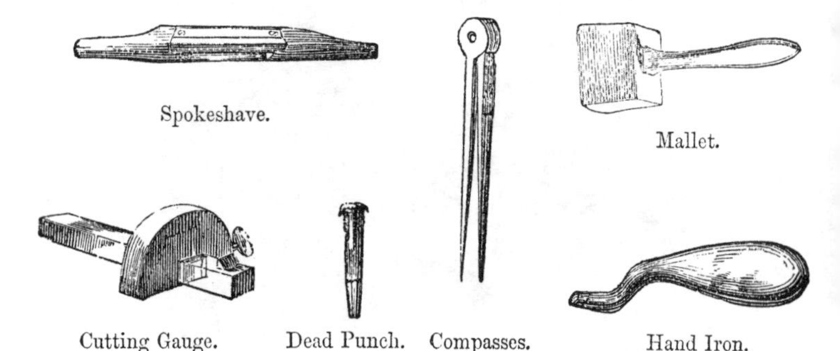

Spokeshave. Mallet.

Cutting Gauge. Dead Punch. Compasses. Hand Iron.

skilful making of saddles. To fit the pigskin over the iron shape, and skilfully to arrange the padding, is often a very troublesome task, where a horse requires some peculiarity of make before he can be well fitted; and to ensure a perfect smoothness and finish a skilful use of the *hand iron* and the *spokeshave* is necessary. Similar care has to be taken in forming the collars to which the harness is attached; and indeed no part of the Saddler's work can be carelessly performed without serious risk· either to the horse or its owner.

As a great part of the Saddlers' work consists of sewing,

he uses the *clamps* to hold the leather between his knees, in the same way as they are used by the shoemaker; but although he employs the *sewing awl* for drilling holes, or, at all events, the *pricking iron*, the sewing is done with *needles* and strongly waxed *thread.* The various kinds of *knives*

Needle and Thread.

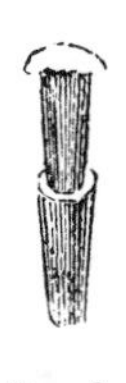

Punch.

Round Knife.

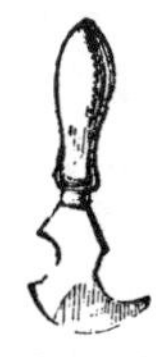

Hand Knife.

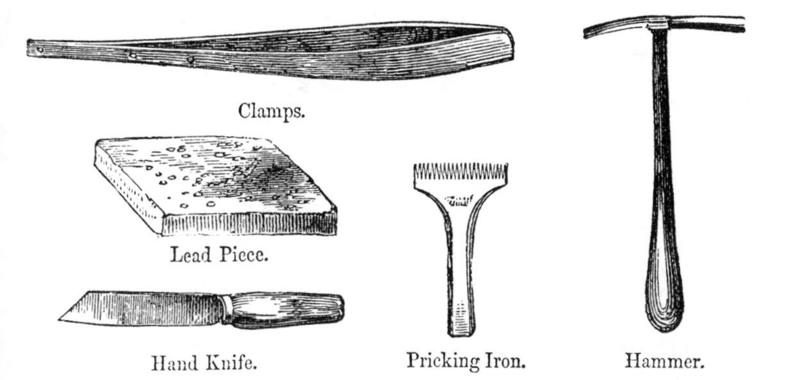

Clamps.

Lead Piece.

Hand Knife.

Pricking Iron.

Hammer.

are used for cutting and paring the leather; the *cutting gauge* and *compasses* for regulating the cutting, the *hammer* for driving the small nails used in the work, and the *mallet* for striking the *pricking iron* or *punches.* The *punch* is used for making the holes in the straps to receive the tongue of the buckle, and is therefore a hollow tube with a sharp cutting edge, so that it will cut out a little round piece of the leather. The *dead punch* is made solid, and is not intended for cutting. The leather, when it has to be cut with the punch, is placed on the *lead piece*, a small square block of lead, which being soft allows a slight yielding of the leather, and at the same time does not blunt the edge of the punch when it has passed through the hole.

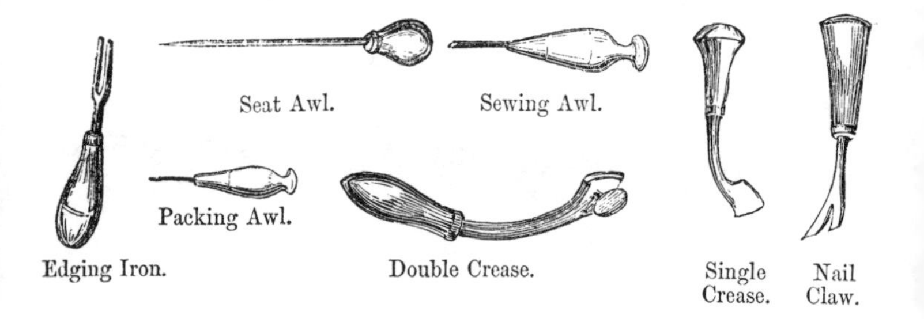

The *seat awl* and *packing awl* are used in the padding and making of saddles and collars; the *nail claw* for removing nails by which the leather has been fastened down.

The various kinds of *creases* are for the purpose of making channels in the leather along the edges which have to be sewn, so that the stitches are sunk below the surface, and the thread will not so easily wear out. The *edging iron* is for a similar purpose. In common saddlery some of the comparatively unimportant straps, or the smaller gear, are not sewn at the edges, and indeed do not require it, although a great deal of the Saddler's sewing is for ornamental purposes. In order to make the whole look uniform, however, these straps are not left plain, but are creased at the edges, and the channels thus made are marked with the *pricking iron*, to give them the appearance of having been stitched. The Saddler is better off than the shoemaker, inasmuch as he generally sits to work at a bench, and need not occupy such a constrained and unhealthy position.

Some account has already been given of the preparation of leather, but it will be desirable here to mention other sorts, some of which particularly belong to the business of harness making.

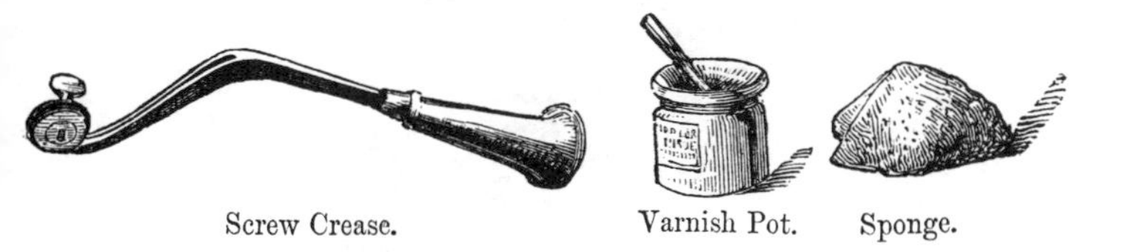

Screw Crease. Varnish Pot. Sponge.

Sheep-skins, when simply tanned, are employed for inferior bookbinding, for leathering bellows, and for various other purposes for which a cheap leather is required. All the *whit-leather*, as it is termed, which is used for whip-lashes, bags, aprons, &c. is of sheep-skin; as are also the cheaper kinds of *wash-leather*, of which gloves, under-waistcoats, and other articles of dress, are made. Mock, or imitation morocco, and most of the other coloured and dyed leathers used for women's and children's shoes, carriage-linings, and the covering of stools, chairs, sofas, writing-tables, &c. are also made of sheep-skin.

Lamb-skins are mostly dressed white or coloured for gloves; and those of goats and kids supply the best qualities of light leather, the former being the material of the best morocco, while kid leather affords the finest material for gloves and ladies' shoes. Leather from goat-skins, ornamented and sometimes gilt, was formerly used as a hanging or covering for walls.

Deer and antelope skins, dressed in oil, are used chiefly for riding breeches. Horse-hides, which, considering their size, are thin, are tanned and curried, and are used by the Harness Maker, especially for collars, and occasionally,

Harness.

when pared thin, for the upper leathers of ladies' walking shoes. Dog-skins are thick and rough, and make excellent leather. Seal-skins produce a leather similar but inferior to that supplied by dog-skins; and hog-skins afford a thin but dense leather, which is used mostly for covering the seats of saddles.

Currying is the general name given to the various operations of dressing leather after the tanning is completed, by which the requisite smoothness, lustre, colour, and suppleness, are imparted. The processes of the Currier are various. The first is styled "dipping" the leather. It consists in moistening with water, and beating upon a trellis-work of wooden spars with a *mallet* or *mace*. After this beating, by which the stiffness of the hide or skin is destroyed, it is laid over an inclined board, and scraped and cleaned, and, wherever it is too thick, pared or shaved down on the flesh side by the careful application of various two-handled knives, and then thrown again into water, and well scoured by rubbing the grain or hair side with pumice-stone, or with a piece of slatey grit, by which means the *bloom*, a whitish matter which is found upon the surface in tanning, is removed.

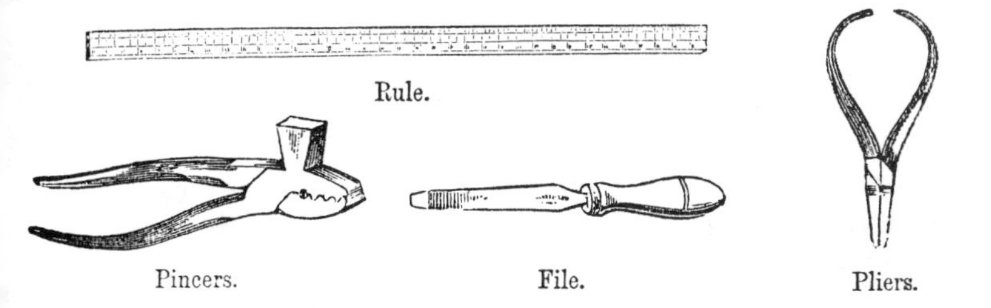

Rule.

Pincers. File. Pliers.

The leather is then rubbed with the *pommel*, a rectangular piece of hard wood, about twelve inches long by five broad, grooved on the under surface, and fastened to the hand. The Currier uses several of these instruments, with grooves of various degrees of fineness, and also, for some purposes, pommels of cork, which are not grooved at all. The object of this rubbing is to give grain and pliancy to the leather. The leather is then scraped with tools applied nearly perpendicular to its surface, and worked forcibly with both hands, to reduce such parts as may yet be left too thick, to a uniform substance. After this it is dressed with the *round knife*, a singular instrument which pares off the coarser fleshy parts of the skin. In addition to these operations, the Currier uses occasionally polishers of smooth wood or glass, for rubbing the surface of the leather; and when the leather is intended for the use of the shoemaker, he applies to it some kind of greasy composition called *dubbing* or *stuffing*.

Leather is occasionally dressed "black on the grain," or having the grain side instead of the flesh coloured. The currying operations in such a case are similar to those above described, but the finishing processes are rather different. The leather is rubbed with a grit-stone, to remove any wrinkles and smooth down the coarse grain. The grain is finally raised by repeatedly rubbing over the surface, in different directions, with the pommel or graining board.

Japanned leather of various kinds is used in coach making, harness making, and for various other purposes. Patent leather is covered with a coat of elastic japan, which gives a surface like polished glass, impermeable to water; and hides prepared in a more perfectly elastic mode of japanning, which will permit folding without cracking the surface, are called enamelled leather. Such leather has the japan annealed, something in the same mode as glass; the hides are laid between blankets, and subjected to the heat of an oven at a peculiar temperature during several hours.

THE HATTER.

CUTTING MACHINE.

THE trade of the Hatter is confined to a few countries of Europe; but, as the fashion of hats is at present mostly restricted, both here and in most parts of the Continent, to those black stiff cylindrical coverings of silk which custom has ordained we shall wear, greatly to our own discomfort, and at a great expense, the business is of considerable importance, and the making of hats has grown into a large branch of industry. Before the introduction of that kind of silk which bears a long pile, like velvet, and is known as *plush*, beaver was the material principally used for hat making; but beaver hats are much more expensive, though they are now seldom made of the skin or fur of the

Cutting out.

Bowing.

beaver, which has grown remarkably scarce, in consequence of the land which the animals formerly occupied having become inhabited.

Beaver hats of the finest quality are made with lamb's wool and the fur of English rabbits. To form the body of the hat, the wool and rabbit's fur are separately *bowed* by bringing a set of strings attached to a bow in contact with a heap of the material, and then striking the strings, so as to cause violent vibrations, and thus separate the filaments.

The two substances are next bowed together until they are intimately mixed; after which the mass is spread evenly, covered with an oil-cloth, and pressed to the state of an imperfectly tangled felt. The next process is to cover the felt with a triangular piece of damp brown paper, and then to fold it in a damp cloth, and work it well with the hand, pressing and bending, rolling and unrolling it, until the interlacing or felting is much more perfect, and the mass is compact. The felt thus prepared is next taken to the wide brim of a boiler charged with hot water and beer grounds, and a small quantity of sulphuric acid; it is well rubbed and rolled until it no longer contracts. The

Bow for Beaver.

felt is next stiffened with shellac, a solution of which is applied by means of a brush to both sides of the felt; after which it is heated in a stove, and by this means the whole substance becomes duly impregnated with the resin; this renders the hat nearly waterproof.

To form the nap of a hat, one half or three-fourths of an ounce of beaver, and some other less costly fur, are bowed together and imperfectly felted in the manner already described, and shaped the same as the body to which it is to be applied; that body is then softened by immersing it in the boiler, when the nap is applied and worked as in felting, until the required union is effected between the two bodies.

The felt thus covered is brought to the proper shape by working it on a wooden block, and is then dyed black.

The hat is softened by steam, the crown is strengthened by placing in it a disc of scaleboard, and linen is pasted over this.

The nap is raised and a uniform direction given to its fibres by means of warm irons and hair brushes. The last processes are lining and binding, when the hat is ready to be worn. In the low-priced hats of the present day, com-

Lathe. Finishing.

moner wool and fur, and smaller quantities of each, are used.

Silk hats consist of a cover or exterior part made of silk plush, which is laid upon a foundation of chip, stiffened linen, or some other light material, previously blocked into shape. The so-called velvet and satin hats deserve those titles only so far as the plush resemble those materials. The plush is mostly woven in the north of England. Paris hats are for the most part made in England, the silk plush being imported from France.

The various tools employed are mostly used for shaping the different parts of the hat. The *cutting machine*, for dividing and preparing the felt and other materials; the *block*, on which the body of the hat is formed previous to

Rounding Card, No. 1.

Rounding Card, No. 2.

Polishing Block.

Finishing Iron.

Blistering Iron.

Multer.

Brow-piece.

Card.

Brush.

its being placed in the *lathe*, where it is made perfectly round; the *brow-piece*, for shaping the hat where it fits the head; the *card*, a sort of wire brush for scratching up or carding the fibre of the silk or beaver after the hat is made; the *rounding cards*, for pressing and completing the hat near the brim; the *curler*, for turning up and shaping the brim; the *multer*, *finishing iron*, and *blistering iron*, are used for shaping and smoothing; the *polishing block* holds the hat while it receives the final process of brushing and smoothing with a pad of velvet called the *veleure*.

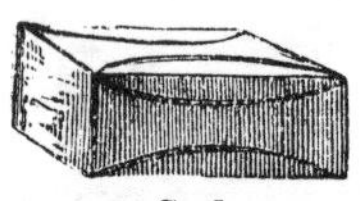
Curler.

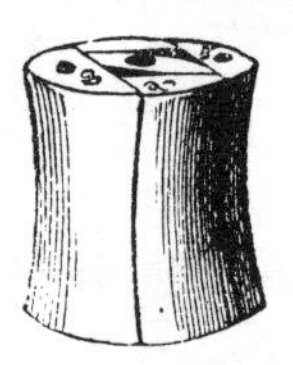
Block.

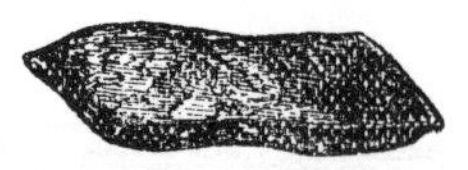
Veleure.

THE MILLER.

WATER-MILL.

THERE is scarcely a boy to be found who, when he has been into the country and seen the wheel of the old water mill going round, or the fans of the windmill slowly revolving in the air, has not thought that it would be pleasant to follow the business of a Miller.

Millers are proverbially jolly fellows, and their houses and the mills themselves are generally very picturesque, and stand in pleasant country places; then, again, what wonderful fat perch and chub and pike can be caught in the weir or the mill stream, and what a quiet sleepy occupation it must be to lie on the grass or in some great room in the mill, watching the fans, or listening to the summer breeze wafting through the sails!

If anybody should think, however, that the Miller's life is a lazy one he had better alter his opinion; for, unless he wishes to starve amidst plenty, the Miller must be up betimes, and, besides working himself, keeping a good look-out amongst his men, lest both he and his customers should suffer by their negligence.

The mortar would seem to be the earliest machine used for the purpose of bruising or reducing grain to a powder, or into a state fit for the making of bread. By means of the handle the pestle would with tolerable facility be driven round the mortar, and the grain reduced to a powder, as is done with certain drugs by the pestle of the apothecary of the present day.

In process of time, shafts were added to these machines; and in the opinion of Beckmann, the oldest cattle mills resembled those described in Sonnerat's Voyage to the East Indies, in which the pestle of a mortar, fastened to a stake driven into the earth, is affixed to a shaft, to which two oxen are yoked. These oxen are driven by a man, while another is employed in dropping the grain into the mortar and placing it under the pestle.

We have good reason for knowing that the Romans, for a long time, used no other instrument than the pestle and mortar.

Pounding continued in use among the Romans so late as after the era of Vespasian. This fact clearly proves that the Romans were many ages behind the Eastern world in the arts of civilization, for grinding of corn into flour was practised, we know, in the times of the patriarchs, and was probably the invention of the antediluvian world.

The subject-Britons universally adopted the Roman name, but applied it, as we their successors apply it at present, only to the Roman mill; still distinguishing their own original mill as we distinguish it, by its own original denomination of a quern. A Roman or water mill was probably erected at every stationary town in the kingdom, and it is quite certain that one was erected at Mancunium (Manchester), serving equally the purposes of the town and the uses of the garrison. One alone would be sufficient, as the use of hand-mills was at that time very common in both, many such having been found about the site of the station particularly, and the use of them generally having being retained among us very nearly to the present period. Such mills it would be particularly necessary to have in the station, that the garrison might be prudently provided against a siege.

The ancient Asiatic hand-mill consists of two flat round stones, about twenty inches or two feet in diameter, kept rolling one on the other by means of a stick, which does the office of a handle. The corn falls down on the undermost stone, through a hole in the middle of the uppermost, which by its circular motion spreads it on the undermost, where it is bruised and reduced to flour; this flour, working out at the rim of the millstones, lights on a board set on purpose to receive it. The bread made of it is said to be better tasted than that made by either wind or water mills: these hand-mills cost only a few shillings of English money.

This description of mill is frequently alluded to in the

Scriptures, and it was the practice for the women to grind corn every morning by means of hand-mills; and in the East, or at least in many parts of it, it continues to be the practice to this day.

Water wheels, as far as the figure and construction are concerned, may be reduced to three kinds, and they are usually known by the names—overshot wheel, balance wheel, breast wheel, and undershot wheel.

Mill.

In an overshot wheel, the water is conducted over the top of the wheel, and acts first by its momentum or mere movement, or motion, and then by its weight—the weight being its principal power in impelling the wheel round, the mere movement or motion of the water producing in this case little effect. The water is received into buckets placed all around the circle of the wheel. It first strikes the wheel at the top, and filling the first bucket, by its momentum or moving power, and more particularly by its weight, it sets the wheel in motion, and consequently makes that side heavier; and as fresh buckets rise to receive the water, while those below have emptied themselves, a constant tendency to motion is created, and rotation is produced.

In a balance wheel, the water strikes the wheel not at the top, but always more or less above its centre, or axle. This wheel differs in no respect from the overshot wheel in its construction. It is employed where there is not a sufficient fall for an overshot wheel, which requires less water in consequence of its commanding a much greater leverage.

In an undershot wheel the water acts only by its momentum, or moving power. The circumference of the wheel, instead of being supplied with buckets, as in the overshot, and breast or balance wheels, is furnished with floats, or float boards, as they are called, and, being exposed to the action of a running stream, generally, if not always, increased in rapidity by making an artificial fall, is thus driven round, and in flour-mills its force is communicated by cog-wheels to the stones employed in reducing corn to meal.

In a breast wheel, the water has no previous fall, and therefore does not strike the floats at the bottom of the wheels, as in the undershot wheel, with an increased velocity. The breast wheel is therefore fixed in what is called a race, formed of stone or brick work, agreeing with the curvature of the wheel, and being thus let on from its own level, acts both by its weight and momentum, or movement. This wheel is unlike the undershot wheel, being close boarded round its circumference, like an overshot or balance wheel, the undershot wheel being always open; in short, it is a sort of bucket wheel; the buckets, however, being constructed differently from those of the overshot and balance wheels.

A tide-mill, as the name imports, is worked by the ebbing and flowing of the tide. Of these mills there are various kinds; first, those in which the water wheel turns one way when the tide is rising, and the other when it is falling; secondly those in which the wheel turns the same way whether it is rising or falling; thirdly, those in which the wheel itself rises or falls as the tide flows or ebbs; and fourthly, those in which the axle of the water wheel is so fixed, that it shall neither rise nor fall; the rotary motion being still given to the wheel, whether it be partially or wholly immersed in the water.

All the machines for grinding corn and seeds are mills, whatever may be their particular application. One very common form is that of an iron machine, supported on four legs, having a winch handle on one side, a fly wheel on another, a hopper at the top, and a crushing apparatus in the centre. The grain or seed is put into the hopper, the winch handle is turned, the grain becomes crushed to powder, and falls out at the bottom of the apparatus. Sometimes the mill is made chiefly of wood, but with iron wheel and crushing apparatus. One kind of wheat mill, in addition to the usual mill apparatus, has a chest which acts as a flour-dressing machine. Some mills are adapted for crushing beans rather than seeds.

The crushing apparatus in mills is of two kinds, either one stone working round in contact with another, or two metallic surfaces, between which the substance is forced, but between which it cannot pass except in a fine state.

The *millstones* employed in grinding corn require to be made of a peculiar kind of stone. The greater proportion of our millstones are procured from a particular spot in Western Germany. At about ten miles from Coblentz is a small town called Andernach, the chief trade of which is in millstones, procured from the neighbouring quarries of Nieder Mendig. There are several quarries, averaging about fifty feet in depth, each quarry shaped like an inverted cone, down the sides of which the quarrymen descend by a spiral path. The quarrymen have to cut away through a superincumbent layer of soft porous stone, till they come to a layer of hard, blackish, heavy stone, regularly porous, and yielding sparks when struck with iron. This is the millstone, and requires good and well-prepared tools to work it; it is supposed to be a compact lava from some extinct volcano; and as there are fissures or gaps at intervals, these facilitate the separation of the stone into blocks suitable for millstones. All round the bottom of the conical cavities, the stone has been excavated in galleries or horizontal passages. The stones are brought to shape by means of hammers and chisels. A deep socket is cut through the middle of such stones as are intended for runners, or upper stones. The furrows on the surfaces of the stones are produced by means of a double-edged hammer, about 14 lbs. weight.

Windmills are of two kinds; in one the wind is made to act upon vanes or sails, generally four, which are disposed so as to revolve by that action in a plane which is nearly vertical; and in the other, the axis of revolution being precisely vertical, any point on the surface of a vane revolves in a horizontal plane. The former is called a *vertical* windmill, and the latter a *horizontal* windmill.

The building for a vertical windmill is generally a wall of timber or brickwork, in the form of a frustrum of a cone, and terminated above by a wooden dome, which is capable of revolving horizontally upon it. A ring of wood, forming the lower part of the dome, rests upon a ring of the same material at the top of the wall, and the surfaces in contact being made very smooth, the dome may easily be turned round upon the wall; and is prevented from sliding off by a rim which projects from it, and descends over the interior circumference of the lower ring. The dome in turning carries with it the windsails and their axle; and thus the windsails may be adjusted to agree with the direction of the wind, or the plane in which the radii of the sails turn may be made perpendicular to that direction. The revolution is sometimes accomplished by the force of a man applied to a winch near the ground, but in general the wind itself is made to turn the dome or the mill by means of a set of small vanes, which are situated at the extremity of a long horizontal arm projecting from the dome, in a plane passing through the vertical shaft of the mill, and on the side opposite to the great sails.

A horizontal windmill is a great cylindrical frame of timber, which is made to revolve about an upright centre, and its convex surface is formed of boards attached in vertical positions to the upper and lower parts of the frame. The whole is enclosed in a fixed cylinder having the same upright centre as the other; this consists of a *revolving screen* or a number of boards, which are so disposed that in whatever direction the wind may blow, it may enter between them on one side only of a vertical plane, passing through the axis, and thus give motion to the interior cylinder. The effective power of the vertical windmill is, however, so much greater than that of the horizontal windmill, that the latter is now seldom constructed.

The *dressing machine* consists of a hollow cylinder, or frame, covered with wire cloth of different degrees of fineness, the finest being at the elevated, or upper end of the cylinder, which is inclined in the same way as the bolting mill. Within the cylinder, which may be made of pieces of wood rendered circular, and, like the ribs of an animal, placed at certain distances from each other, a reel is placed with its axle in the centre of the cylinder, which is fixed, or

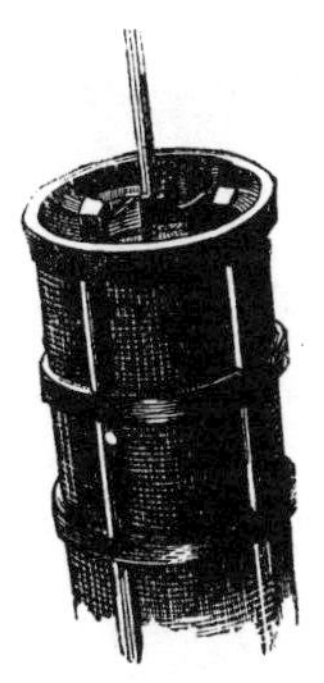
Dressing Machine or Bolter.

Jack.

stationary. To the rails of this reel are attached hair brushes, which, when made to revolve, or turn round, brush against the interior cloth wire surface of the cylinder. The machine is provided with a shoe or jigger, very similar to that of the millstones, to cause a regular supply of flour or meal in the same state in which it came from the millstones through a spout from the floor above, to which it is elevated by the sack tackle, or elevator, after being ground. The meal by this means being gradually let, or fed into the cylinder, is, by the motion of the brushes of the reel, sifted or rubbed through the cloth wire with which the cylinder is covered. The finest of the flour will go through the upper end, where the finest wire cloth is placed; the next finest through the next division of the wire cloth, which is coarser, the middlings through the following division, the pollard, or sharps, through the last, and the bran, not being able to get

Revolving Screen.

Jacob's Ladder.

through the wire cloth at all, on account of its coarseness, is thrown out at the end of the cylinder. The cylinder is enclosed in a large and close box, to prevent the waste of flour by its going off in dust. This box is divided into several compartments, or partitions, by means of moveable boards. Some millers have more partitions than others, and indeed they all vary the number to suit the nature or sort of flour which they are manufacturing. In a dressing machine of three divisions, the flour deposited in the first is called household, or seconds, that in the second middlings, and in the third pollard, which is not flour, but a fine description of bran.

When wheat arrives at a mill to be ground the sacks are received in the lower part of the mill, and hoisted by means of the sack tackle to the upper storeys, generally the uppermost. The mode of this operation is as follows, and it is performed with little bodily labour. The rope or chain of the sack tackle is firmly fastened round the mouth of the sack by the man below, who by means of a rope attached to a lever throws the tackle into gear. The sack then immediately ascends, without any further aid from the man, through the different trap doors, till it has arrived at the place of its destination. There another man is ready to receive it, who, as soon as he has landed it by pulling it on one side, throws the tackle out of gear, and returns the rope or chain to the man below. The wheat is then shot into a garner or bin, and thus the same process goes on till the whole load or cargo is safely deposited in the bin or garner. The wheat remains there till it is wanted to be ground, when, by means of a spout, it is conveyed to the hoppers below, and from thence runs in between the stones. In its progress to the stones, it may, or may not, be subjected to a cleansing process. The wheat being reduced to a flour, escapes through an aperture of the floor into a spout, by which it is conveyed to the trough. It is then either put into sacks, and drawn up again into one of the higher storeys, and deposited in bins over the dressing machines, or this is effected by a machine called an *elevator*, which performs the operation without the assistance of manual labour. In this bin the flour should be left till, at any rate, it is perfectly cold, but is will be all the better if it remain for three or four weeks, provided it be occasionally turned. From this bin it is passed, by means of a spout, to the hopper of the bolting mill or dressing machine, by which it is separated from the bran or pollard, and is then fit for use.

Smutter. Scoop. Claw.

The *screening machine* consists of a roller-shaped sieve, so divided, that the corn which is placed at one end passes over a large surface of wire as the sieve revolves, and this operation removes from it external impurities, such as sand and dirt. When it arrives at the end of the screw the wheat falls into a hopper, by which it is conveyed by spouts into small hoppers, placed over one side of each pair of millstones. From these hoppers there are spouts placed in nearly a horizontal line, which spouts conduct the corn from the hoppers to the eye of the millstone. They are attached to the hopper, so as to admit of a horizontal motion, which is effected by a projecting part of the axle or spindle of the stone, called a *damsel*. It is shaped like

Scales and Weights.

Troughs.

a cross, so that, by the revolution of the millstone, it keeps tapping the end of the spout, and gives a rapid vibratory motion to it, which causes a regular supply of corn to enter the eye of the stone; by the action of the stones the corn is reduced to flour, which passes, by means of spouts, into the troughs on the ground floor.

The *troughs* in which the flour is kept; the *scoop* and *scales* for weighing small quantities; the *claw* for moving and the *steelyard* for weighing sacks, are the other principal objects seen in a mill. *Jacob's ladder* is the name given to a revolving band fitted with a number of leathern cups. These revolve with the band through the flour and serve to carry it up a shaft from one floor to another.

Steelyard.

THE BAKER.

BAKING OVEN AND KNEADING TROUGH.

Of all the trades that are carried on in large towns there is none more important than that of the Baker. In some parts of the country, where people make their own bread at home, or at all events have all the materials for making it, and know how to mix the different ingredients, it is of less consequence; but only imagine what would be our dismay in London, if we got up some morning and heard that all the bakers had agreed not to send in the rolls for breakfast, and that we must be satisfied to live upon puddings and vegetables until they set to work again to make bread.

And yet, in the early part of their history, several nations of which we read at school had no knowledge of the trade of a Baker. Until they discovered the art of making proper bread, and somebody showed them the use of an oven, the Romans made their meal into a sort of porridge, or knew no better than to mix it into flat cakes, which they cooked on hot stones or in the wood ashes of their fires. The Anglo-Saxons were a little better off, for they mixed leaven with their dough, to make the cakes lighter and better; but you will remember—by the story of King Alfred, who when he was in disguise and going to watch the Danish camp, was left in the neatherd's hut to watch the cooking of the cakes—that they were baked in the embers of burnt wood upon the hearth.

In many parts of Europe they still use a sort of cake-bread or rolls instead of loaves; and in some parts of northern Germany and Russia the bread of the peasants is nearly black, and is so coarse and sour that few English people could eat it, especially as it is kept till it grows quite hard, and sometimes has to be cut with a saw.

The Eastern nations understood the use of ovens, however, and the Jews especially were very good Bakers, as were probably the Egyptians, for we read of Pharaoh's Baker, whose dream Joseph interpreted. The trade of baking was held in very high esteem amongst the nations of antiquity, and included not only the making of bread, but of those cakes and sweetmeats of which all Oriental people, and some European people too, are so fond. By reading the history of the Jews and the Egyptians we learn that fermented or leavened bread was in common use amongst them, at a very early period of their history, so that the pulse-eating people, or the people of Europe who could only eat grain made into porridge, or simply cooked, were for a long time behind the Orientals in this respect. It was not till 600 years after the establishment of the Roman state that a public bakery was opened in Rome itself, and before this time, which was about 167 years before the birth of Christ, all the baking and bread making was done, amongst the rest of the family cooking, in the kitchens of private houses.

When the public bakeries were established, however, those who followed the trade were held in great respect, and a code of laws was made to regulate the manner in which their business should be conducted. The same importance was given to the trade in England, when Bakers first set up business in the large towns, and people began to buy their loaves instead of making them at home. The early statutes and laws place Bakers above mere handicraftsmen, and ranked them with gentlemen, and very severe punishments were inflicted on fraudulent Bakers, who neglected to mark their loaves so that "wheaten" bread might be distinguished from "household;" the same laws condemned that Baker to the pillory who gave short weight, even by so much as the fraction of an ounce, and the bakehouses were placed under the control of the magistrates.

There are many people who think that some such stringent laws might be usefully employed now, and there can be no doubt that the Baker has too many opportunities of adulterating his bread, or sending short weight: the first seriously injuring the health, and the second the pockets, of his customers. One of the worst features of the Baker's trade in our time, however, is the dirty condition of the bakehouses and places where bread is made, the filthy habits which such places give rise to, and the very long hours during which journeymen Bakers are at work. All these matters have lately been made the subject of inquiry, and it may be hoped that they will be greatly improved.

Egg Whisk. Flour Basket and Scoop. Egg Brush.

The tools that are used in most bakeries are, beside the *oven*, where the bread is baked, and the *kneading trough*, in which the dough is mixed: a *seasoning tub* for mixing other ingredients to be mingled with the dough, a *wire sieve* for sifting the flour, and a *seasoning sieve* made of tin pierced with small holes; a *flour basket* lined with tin, and a *flour scoop*, a *pail*, a *bowl*, a *salt-bin*, which should be near the oven, a *yeast tub*, a *dough knife*, *scales* and *weights* for weighing the dough before it is moulded into a loaf, a *scraper* for removing the dough from the trough and the board where the loaves are shaped, wooden and iron *peels*, a

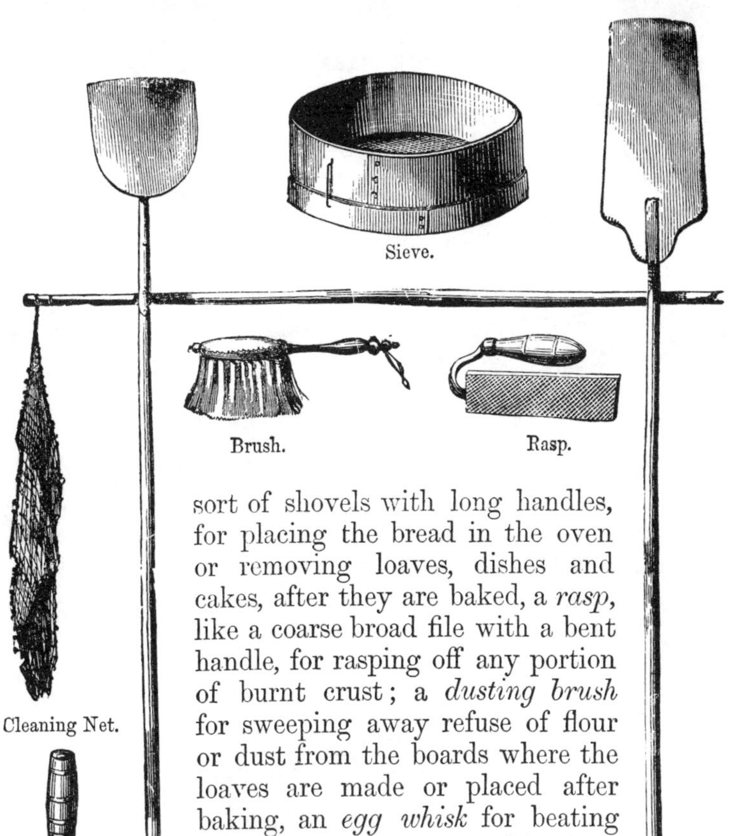

sort of shovels with long handles, for placing the bread in the oven or removing loaves, dishes and cakes, after they are baked, a *rasp*, like a coarse broad file with a bent handle, for rasping off any portion of burnt crust; a *dusting brush* for sweeping away refuse of flour or dust from the boards where the loaves are made or placed after baking, an *egg whisk* for beating eggs used in pastry, and an *egg brush* for putting a glazing of egg on the outside of buns or cakes,

Biscuit Marker.

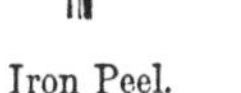
Paste Cutter.

Knife. Iron Peel. Wooden Peel.

differently shaped *tins* or *moulds* for rolls or other articles of fancy bakery, coarse squares of *baize* or *flannel* for covering the dough or the newly-made bread, and a *scuttle, swabber,* or *cleaning net,* made of a quantity of rough netting fastened on the end of a pole, and which,

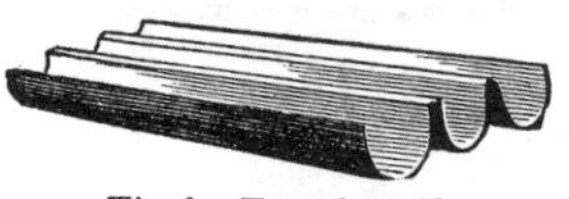

Tin for French Rolls.

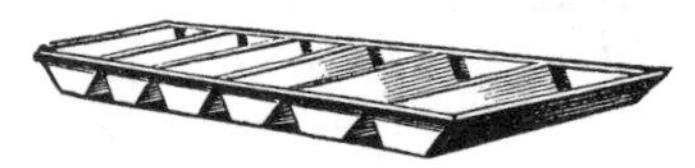

Tin for Sponge Cakes.

after being wetted, is used for the final removal of all dirt from the oven just before "setting the batch," or placing the loaves for baking. Beside these there are in most bakehouses *set ups,* or oblong pieces of beech wood, to be placed in the oven for the purpose of keeping the loaves in their places.

Scales and Weights.

It would be impossible to give instructions here how to make the various sorts of bread, but the ordinary kind sold by London Bakers is made much in the following way :—

Suppose that the Baker desires to make up a sack of flour; he empties it into the kneading trough, and then proceeds to sift it, in order to make it lie more lightly and to break up any lumps.

He then takes from eight to ten pounds of potatoes and boils them, without removing the skins, afterwards mashing them in the seasoning tub, with about half the quantity of flour, and adding a couple of pails of water. To this mixture he pours about two quarts of yeast, made from the liquor of boiled hops, malt, and patent yeast already made and sold for the purpose. This is the leaven, which makes the difference between bread and meal-cake, or the English loaf and the Australian "damper," which is made only of flour and water, and baked at a bush fire; the yeast or leaven is, in fact, meal in its early state of decomposition.

This mixture then is left in the seasoning tub, covered with a sack for several hours, and allowed to ferment. Then the Baker separates about a quarter of the flour in the trough by means of a board, and piles it up at one end apart from the larger quantity, and it is upon this that he pours the contents of the seasoning tub, taking care that it passes through a sieve placed on a couple of sticks across the trough.

This portion upon which the liquor has been poured is called the *sponge*, and he proceeds to "set it" by thoroughly mixing, and finally giving it a dust of flour at the top, after which it is kept for five or six hours, or until it has twice "risen," or puffed out by means of the fermenting liquor within it.

Just as it has risen a second time, and air bubbles are breaking through it to the surface, about three more pailfuls of water are poured upon it, and in this water about three pounds of salt have been dissolved; this is well mixed with the sponge, and then, the board being removed, the sponge and the rest of the flour is worked into one mass. It is this kneading and breaking up the sponge which is the hard labour of the Baker, and that it is hard labour may be known from the fact that he works nearly naked, and that, as he lifts and pummels the tenacious mass, he heaves great sighs and groans like those with which paviors ram down the stones in the roads. In many of the best bakehouses now a machine is used, which supersedes this manual labour; the mass of dough being placed in a cylinder, within which an axle fitted with bent blades, or arms of iron, revolves, passing through and through the dough as it moves from end to end.

By another system (that of Dr. Dauglish) the use of yeast is dispensed with, and the bread, made by machinery, is leavened by carbonic acid gas, which is forced into the cylinder after the atmospheric air is pumped out, and so goes through the dough and produces the results of fermentation.

The dough having been thoroughly mixed in the trough, is left for an hour or two to "prove," and then, after being sprinkled with flour to prevent its sticking, is thrown out upon a board, or the lid of the next trough, and cut into pieces which are weighed, and afterwards moulded into loaves. The moulding is only learnt by practice. The piece of dough is cut in half and shaped according to the kind of loaf required, one piece forming the top of the loaf, being laid in a hollow of that which is to be the bottom, and the joint made by a skilful turn of the knuckles. The loaves are placed in the oven by means of the peel, and are packed at the back and sides as closely as possible, the cottage bread only being separate, that it may be crusted all round. The batch takes about two hours to bake. Biscuits are now mostly made by machinery, which in large establishments turns them out ready for the oven; but when they are wholly or partially made by hand, the dough is prepared and afterwards moulded into shapes, and each shape pricked with the *docker*, before being placed in the oven.

Cornfield.

THE SUGAR REFINER.

BOILING HOUSE.

ALTHOUGH sugar was known from very early times, it was used only in medicine, and was supposed to be a sort of honey found upon canes in India and Arabia. It is frequently mentioned by the very early writers. The culture of the canes seems to have been confined to the islands of the Indian Archipelago, and the kingdoms of Bengal, Siam, &c. The traffic in sugar was so lucrative that the Indians concealed the mode of preparing it, stating to the merchants of Ormus, who imported it with gums and spices, that it was extracted from a reed, whereupon many unsuccessful attempts were made to find it in the reed-like plants of Arabia. In 1250 the great discoverer, Marco Polo, visited the country of the sugar cane, and the merchants afterwards sent to the place of its growth, instead of buying it at Ormus. For a long period the use of sugar in England was confined to medicines, or to preparing choice dishes at feasts; and this continued till 1580, when it was brought from Brazil to Portugal, and thence to our country.

Cultivators distinguish three great varieties of canes—the Creole, the Batavian, and the Otaheite. The Creole cane is indigenous to India, and was transplanted thence to Sicily, the Canary Isles, the Antilles, South America, and to the West Indies. It has dark green leaves, and a thin but very knotty stem. The Batavian or striped cane, which has a dense foliage, and is covered with purpled stripes, is a native of Java, where it is chiefly cultivated for the manufacture of rum; it is also met with in some parts of the New World and the West Indies. The Otaheite variety grows most luxuriantly, is the most juicy, and yields the largest product. It is cultivated chiefly in the West Indies and South America; it ripens in ten months, and is hardier than the other varieties.

The sugar cane, being originally a bog-plant requires a moist, nutritive soil, and a hot tropical or sub-tropical climate. It is propagated by slips or pieces of the stem, with buds on them, and about two feet long. It arrives at maturity in twelve or sixteen months, according to the temperature; the leaves fall off towards the following season, and the stem acquires a straw-yellow colour. The cane is cut by some planters before the flowering season, but it is more usual to cut it some weeks after. The plantations are so arranged that the various divisions of the fields may ripen in succession. The land should be supplied with manure rich in nitrogen, but not containing much saline matter. After the harvest the roots strike again, and produce a fresh crop of canes; but in about six years they require to be removed.

The time for cutting the canes varies with the soil and season, and the different varieties of cane. In a state of maturity the canes are from six feet to fifteen feet in length, and from one and a half inch to two inches in diameter. The usual signs of maturity are a dry, smooth, brittle skin, a heavy cane, a grey pith, and sweet and glutinous juice. Canes should be cut in dry weather, or the juice will be found diluted with an excess of water. When cut they are tied up in bundles, and conveyed to the crushing mill, particular attention being paid that the supply should not exceed the demand, otherwise the cut canes would ferment and spoil.

The sugar cane grows from pieces or slips of itself, containing germs, and these develop rootlets at the joints, which draw sustenance to the young shoot as it increases. In the course of time the buds in the radicle, or root-joints of the first cane, throw out roots, and form a radicle for a second stem; and in this way, under favourable circumstances, several canes are produced from the parent stock for a period of about six years, and sometimes for several more. They, however, diminish every year in length of joint and circumference, and are inferior in appearance to the original plant; but they yield richer juice, and produce finer sugar.

The sugar exists in the cells of the cane in a state of solution, and is extracted therefrom by pressure. With this, as with other branches of industry, science has of late years stepped in, and has greatly facilitated the process of extraction and manufacture. It may be as well to give here a summary of the processes employed in the preparation

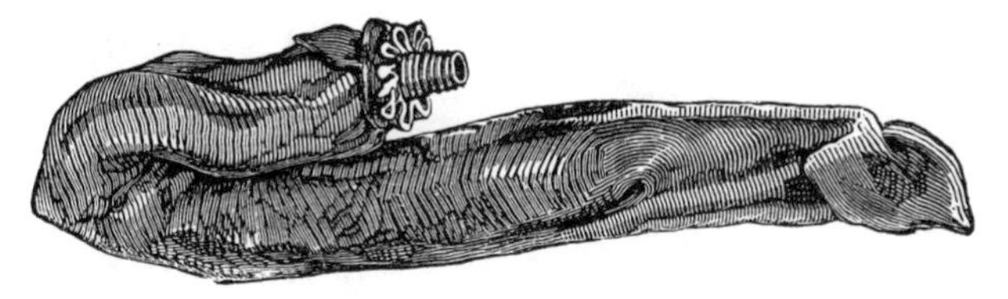

Filtering Bag.

of raw sugar: The canes are passed through the mill, and the juice thus extracted from them runs from the mill into a tank, whence it is pumped to cisterns for supplying the *clarifiers*, heated by steam, where it is purified. From thence it is run into *bag filters*, by which the mechanical impurities are removed. It is then run into *charcoal filters* to remove the colouring matter of the juice. The filtered juice is then run off into tanks and is drawn thence by vacuum into the *vacuum pan*, where it is granulated, and from whence it is finally discharged for packing. When the steam clarifiers are not employed the cane juice is run into a series of pans or *teaches* over open fires. This apparatus is also known as a "battery," and forms another method of purification.

The original crushing apparatus of India was a kind of squeezing mortar, made out of the hollow trunk of a tamarind tree, and worked by a yoke of oxen, the pestle or stamper being a strong beam eighteen feet long, and rounded at the bottom so as to squeeze or crush the canes in the mortar. Mills similar to those used for crushing oil seeds

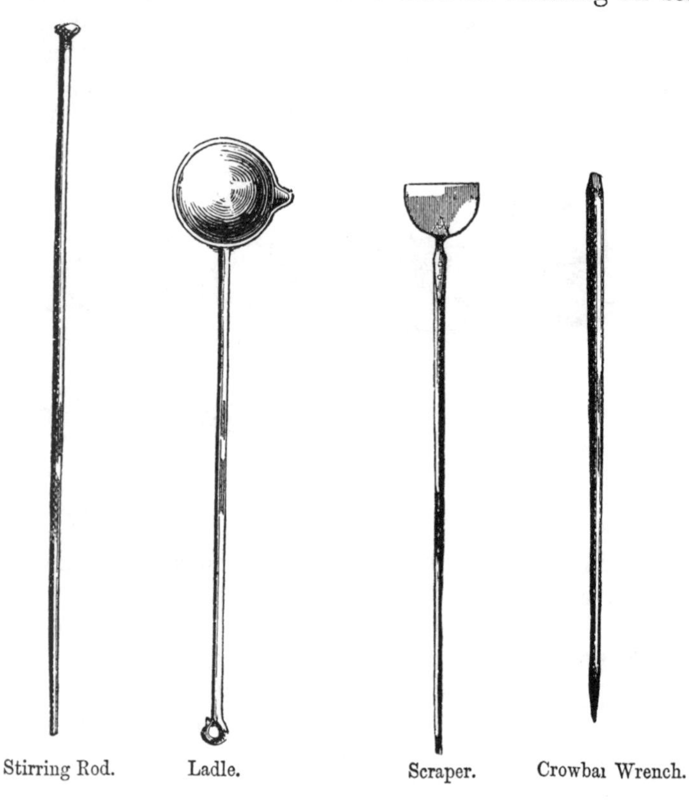

Stirring Rod. Ladle. Scraper. Crowbar Wrench.

were used—as were also several other forms of apparatus —before rollers were introduced. Stone and iron rollers were first used, with the axes in a vertical position, but the horizontal was soon found to be the more convenient and economical. The importance of a systematic mechanical arrangement appears to have awakened attention during the reign of Charles the Second, for in 1663, Lord Willoughby and Lawrence Hyde (second son of Edward, Earl of Clarendon, High Chancellor of England) associated themselves with one David de Marcato, an inventor, and obtained a patent for twenty-one years for making and framing sugar mills. In 1691, John Tizack patented an engine for milling sugar canes, &c.; but in this, as in the previous case, it is not specified how the mill should be made. Later on in 1721, William Harding, a smith, of London, who had been many years in Jamaica, and was skilled in the manufacture of sugar mills, having observed their imperfections, how that they were chiefly made with large wooden cogs cased with iron, endeavoured while abroad to improve their construction, but failed for want of competent workmen. On his return to London he made models of sugar mills which were approved by the Royal Society. These mills were fitted with cast iron rollers and cog wheel gearing, and were worked by water power; from description they appear to be the type of mills of the present day. Some forty-five years later, Yonge and Barclay, ironfounders, of Allhallows-lane, City, applied friction wheels to sugar mills. In 1773, John Fleming, a mill carpenter, proposed an arrangement of windmill sails which turned a vertical timber shaft shod with iron, and which gave motion to two hard wooden rollers, between which the cane was guided and squeezed. In 1807, H. C. Newman, of St. Christopher, West Indies, designed a mill to be worked by horse power. He used cog and crown wheels to give motion to three upright rollers, and the arrangement was considered one which greatly augmented the power and execution of this class of machinery. In 1821, John Collinge, of Lambeth, improved cast iron sugar mills by casting the rollers on wrought iron shafts, instead of keying them on, as previously done. In 1840,

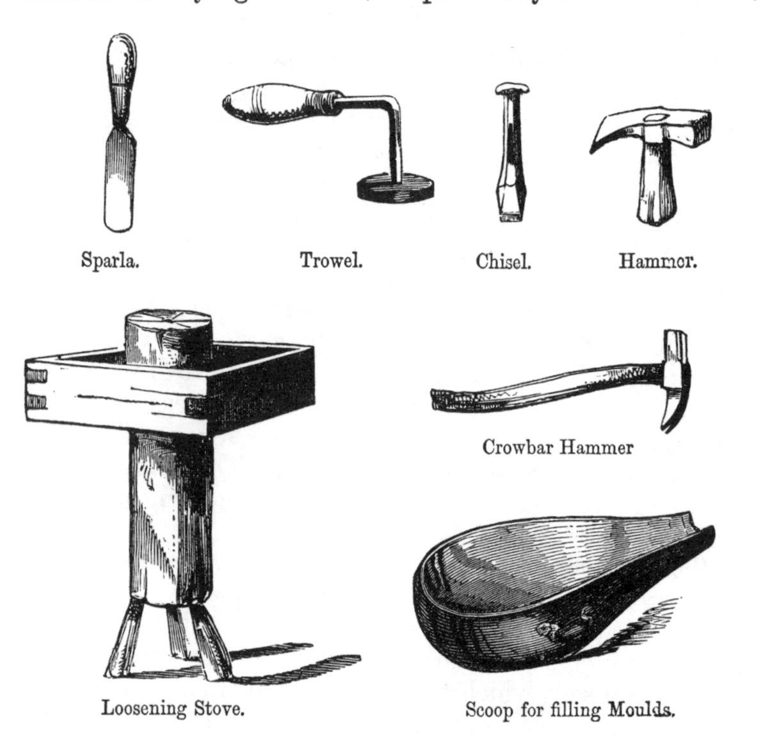

Sparla. Trowel. Chisel. Hammer.

Crowbar Hammer

Loosening Stove. Scoop for filling Moulds.

James Robinson patented improvements in sugar mills, which consisted in using four rollers, one large one and three smaller ones beneath, placed horizontally, and gearing by cogs into each other. Up to that time three rollers only appear to have been used. He also proposed to use six rollers, which are fed from an endless band passing over the rollers. He cast the rollers and shafts in one piece, cored out to admit steam to facilitate the extraction of the juice from the cane during crushing. He also proposed to tin the interior of vacuum pans, &c. Various other arrangements have been patented, but it is unnecessary here to enumerate, much less to describe, them. The foregoing examples give an idea of the progress of the subject during nearly two hundred years. The last ten years have seen rapid strides made in improving the make of mills, and the general arrangement of sugar works.

Cane juice, as expressed by the mill, is an opaque, slightly viscid fluid, of a dull grey, olive, or olive green colour, and of a sweet balmy taste. The juice is so exceedingly fermentable that, in the climate of the West Indies, it would often run into the acetous fermentation in twenty minutes after leaving the mill, if the process of clarifying were not immediately commenced.

The processes followed in the West Indies for separating the sugar from the juice are as follows: The juice is conducted by channels from the mill to large flat-bottomed *clarifiers*, which contain from three hundred to a thousand gallons each. When the clarifier is filled with juice, a little slaked lime is added to it; and when the liquor in the clarifier becomes hot by a fire underneath, the solid portions of the cane juice coagulate, and are thrown up in the form of scum. The clarified juice, which is bright, clear, and of a yellow wine colour, is transferred to the largest of a series of evaporating coppers or pans, three or more in number, in which it is reduced in bulk by boiling; it is transferred from one pan to another, and heated until the sugar is brought to the state of a soft mass of crystals imbedded in molasses—a thick, viscid, and uncrystallizable fluid. The soft concrete sugar is removed from the coolers into a range of casks, in which the molasses gradually drains from the crystalline portion, percolating through spongy plantain stalks placed in a hole at the bottom of each cask, which act as so many drains to convey the liquid to a large cistern beneath. With sugar of average quality, three or four weeks is sufficient for this purpose. The liquid

Besom. Lump Mould. Loaf Mould.

portion constitutes molasses, which is employed to make rum. The crystallized portion is packed in hogsheads for shipment as raw, brown, or muscovado sugar; and in this state it is commonly exported from our West Indian colonies.

The refining of sugar is mainly a bleaching process, conducted on a large scale in England. There are two varieties produced by this bleaching, viz. clayed and loaf sugar. For clayed sugar, the sugar is removed from the coolers into conical earthen moulds called formes, each of which has a small hole at the apex. These holes being stopped up, the forms are placed apex downwards in other earthen vessels. The syrup, after being stirred round, is left for from fifteen to twenty hours to crystallize. The plugs are then withdrawn, to let out the uncrystallized syrup; and the base of the crystallized loaf being removed, the forme is filled up with pulverized white sugar.

This is well pressed down, and then a quantity of clay mixed with water is placed upon the sugar, the formes being put into fresh empty pots.

The moisture from the clay, filtering through the sugar, carries with it a portion of the colouring matter, which is more soluble than the crystals themselves. By a repetition of this process, the sugar attains nearly a white colour, and is then dried and crushed for sale.

But loaf sugar is the kind most usually produced by the refining processes. The brown sugar is dissolved with hot water, and then filtered through canvas bags, from which it exudes as a clear, transparent, though reddish syrup. The removal of this red tinge is effected by filtering the syrup through a mass of powdered charcoal, and we have then a perfectly transparent colourless liquid.

In the evaporation or concentration of the clarified syrup, which forms the next part of the refining process, the boiling is effected (under the admirable system introduced by Mr. Howard) in a vacuum, at a temperature of about 140° Fahrenheit. The sugar pan is a large copper vessel, with arrangements for extracting the air, admitting the syrup, admitting steam pipes, and draining off the sugar when concentrated. In using the pan a quantity of syrup is admitted, and an air pump is set to work to extract all the air from the pan, in order that the contents may boil at a low temperature. The evaporation proceeds, and when completed the evaporated syrup flows out of the pan through a pipe into an open vessel beneath, called the *granulating vessel*, around which steam circulates, and within which the syrup is brought to a partially crystallized state. From the granulators the syrup or sugar is transferred into moulds of a conical form, which were formerly made of coarse pottery, but are now usually of iron. In these moulds the sugar whitens and crystallizes, the remaining uncrystallized syrup flowing out at an opening at the bottom of the moulds. This syrup is reboiled with raw sugar, so as to yield an inferior quality of sugar; and, when all the crystallizable matter has been extracted from it, the remainder is sold as treacle. The loaves of sugar, after a few finishing processes, are ready for sale.

THE DYER AND SCOURER.

DYERS AT WORK.

THE trade of the Dyer may be placed amongst the most ancient of the arts, and the tools that are used in it are so few that they need scarcely any description, since almost all the apparatus required are the various coppers, vats, and other vessels used for boiling the fabrics to be dyed, and immersing them completely in the liquors which have been prepared from the dye stuffs, the proper use of which is the principal secret in the business. With regard to the scouring or cleaning of fabrics, whether made into garments or not, the operation consists in applying detergents, or substances like soap and turpentine, for removing grease and dirt, or other liquids which have a detergent property but will not injure the colours; the use of these, with a *sponge* and a *hard brush*, is nearly all that is required; the

Sponge.

Drying Room.

Hard Brush.

garment sent to be cleaned being stretched on a *frame* and dried, either in the dyer's *drying room*, or in some open situation, where they are least likely to be soiled or spotted.

The art of dyeing was practised to a very great extent by the ancient Egyptians, Phœnicians, Greeks, and Romans; and the island of Tyre was, in very early times, celebrated for a purple dye which was, perhaps from its costliness, used for colouring the robes of kings and emperors, from which practice purple became the imperial colour.

The art of modern dyeing very greatly consists in the proper use of what are called mordants, that is, substances which, although they do not of themselves produce colour, so act upon the dyes as to cause them to give an intenser hue to the fabric, and also serve to make the colour permanent. The modern dyers have obtained several dye drugs unknown to the ancients, such as cochineal, quercitron, Brazil and logwood, arnotts, and indigo, which was

Garment Frame.

only known to the Romans as a paint; but the vast superiority of our dyes must be principally ascribed to the employment of alum and solution of tin mixed with other substances as mordants, which give depth, durability, and lustre to the colours. Another improvement in dyeing is the application of metallic compounds, such as Prussian blue, chrome yellow, and manganese brown, to textile or woven fabrics.

Our readers will see from what has been said of mordants how what are called fast colours are obtained, fast colours meaning colours which will not be affected even by the liquor of the dye bath. Another very necessary subject of information in relation to dyeing is the fact that different substances, such as silk and wool, will not be equally affected by the dye in which they are placed, since the particles contained in the composition of these substances have different degrees of what is called affinity, that is, they combine

Vats and Copper Pan.

Wringing Machine.

in a greater or less degree with the component matter of the dye stuff.

We have already said something of the preparation of cotton. The operations to which silk and wool are subjected before being dyed are intended to separate superfluous substances from the animal fibre, and to make that fibre more easily unite with the colouring particles.

Silk is scoured by means of being boiled in water and soap, whereby the animal varnish is removed from the surface; if intended to be very white, it is bleached by humid sulphurous acid. Wool is first washed in running water to separate its coarse impurities, and is then freed from the greasy animal matter secreted from the skin of the

Dipping Copper, Winch, Punching, &c.

sheep by means of ammoniacal liquor, soap and water, or a solution of soda. It is finally bleached by the fumes of burning sulphur, or by aqueous sulphuric acid.

What are called tinctorial colours, as distinguished from mordants, are either simple or compound. The simple are black, brown or dun, blue, yellow, and red; the compounds are grey, purple, green, orange, and others.

Gall nuts, pyrolignite of iron, logwood, copperas, and verdigris, are the chief materials for producing black. Walnuts, sumach, madder, cochineal, cudbear, acetate of iron, catechu, Brazil wood, arnotts, are all employed in producing brown. Indigo, Prussian blue, and woad for blue. Fustic, Persian

Carboy for Spirits. Syphon. Puncheon. Pole for Stirring.

berries, quercitron, turmeric, and weld for yellow. Cudbear, Brazil wood, cochineal, kermes, lac, logwood, madder, safflower, for red; and various compounds for purple, green, orange, &c.

The dye materials imported from foreign countries are principally cochineal, fustic, gum arabic, gum senegal, gum animi, gum copal, gum tragacanth, indigo, lac dye, shellac, logwood, madder, smalts, valonia, yellow berries, and zaffre.

The tools shown in the engravings explain themselves, and, as has been already said, principally consist of the various coppers and vats, the *syphons* or tubes by which liquor will flow from one vessel to another, and the *puncheons*, for stirring and thoroughly immersing the articles in the dye stuff.

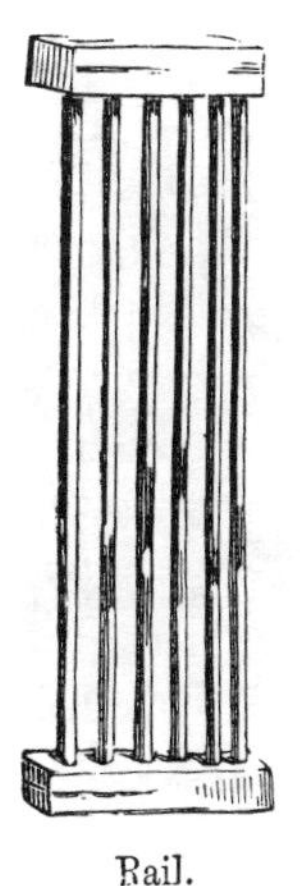

Rail.

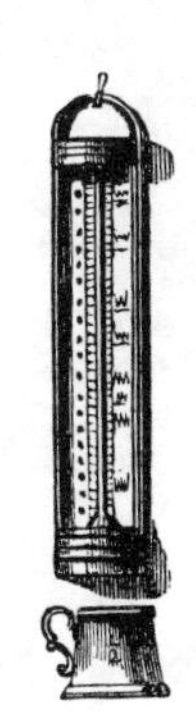

Thermometer and Measure.

Measure.

THE COPPERSMITH.

WORKSHOP.

THE very great variety of purposes for which the metal called copper is used, renders it one of the most valuable productions of this country, and several very important manufactures would be incomplete without it, since it is the principal material of which many large vessels are formed, and is also largely employed as an alloy for other metals. Although the metal is sometimes found in a pure or native state, its most abundant ore is that known as copper pyrites, of which there are many kinds. The most common yellow copper ore is a very abundant mineral, found in large quantities in Cornwall, Devon, and the Isle of Anglesea, and is a compound of about equal parts of copper, iron, and sulphur. The mines of Devon and Cornwall yield more than three-fourths of the copper obtained in England, or about 190,000 tons of ore in a year, the value of which is above a million sterling. The amount of copper annually obtained in the United Kingdom is about 15,000 tons, worth more than a million and a half of money. About the same quantity is imported into this country from Chili and Cuba, and a very valuable copper ore is also found at the Burra Burra mines in Australia.

The copper ores of Cornwall and other parts of the country are generally shipped to Swansea, where coal is abundant, in order to be separated from the metal. For this purpose the ore is heated to redness, or roasted in an open furnace, in order to burn away the sulphur, and the fumes given out by this process are most pernicious. The ore is afterwards melted several times to separate the other impurities, which, when fused, float like scum on the surface of the liquid metal, and are then easily removed.

Copper has a peculiar reddish colour, and will bear a brilliant polish; its smell and taste are both disagreeable. It is one of the most malleable of metals, and can be so readily worked by the hammer that it is beaten out into thin leaves, which, under the name of "Dutch metal," are employed in ornamenting toys, &c. in imitation of gilding.

It is also so ductile that it can be drawn out into finer wires than any ordinary metal, except gold, silver, and iron, and its tenacity is so great that a wire one-tenth of an inch in diameter will bear a weight of 175 lbs. This wire is very flexible, and not very elastic, but when rolled into sheets copper is one of the most elastic, and when struck one of the most vibratory and loud sounding metals.

If taken into the animal system all preparations of copper are violent poisons; and, as this metal is directly acted upon by vinegar and other acids, it should not be employed for making vessels used in cooking or preparing food.

The easy malleability of copper allows it to be rolled into thin sheets, which can easily be hammered into any form that may be desired, and it is this operation which belongs to the trade of the Coppersmith.

Sheet copper is employed in covering the bottoms of ships, to protect them from the attacks of marine animals; it is also used for coins, which are punched out of the sheet of metal and stamped with dies, and for plates on which pictures are engraved; the metal being soft enough to yield to the tool of the engraver, but yet sufficiently hard to resist the pressure necessary to print the picture.

As an ingredient in alloys copper is most valuable, especially in bell metal, which is composed of three parts of copper and one part of tin; in bronze, which is nine-tenths copper and one-tenth tin; and in German silver, argentine, nickel silver, and other alloys used for making forks, spoons, dishes, &c. which are composed of copper, zinc, and nickel. These are very beautiful alloys, closely resembling silver; but the copper they contain is liable to be partially dissolved by any long exposure to the action of acids.

Copper is used in small quantities as an alloy for gold and silver both in coinage and in plate, to give the requisite degree of hardness. Gold used for coins, or what is called standard gold, is formed of eleven parts of pure gold and one part of copper. Being a better conductor of electricity than any other metal, it is largely employed in the formation of telegraphic wires; and, as it is not hard enough to strike fire with flint or grit, it is used in gunpowder mills and magazines instead of iron.

The processes of casting and rolling copper into sheets are so much like those already described in the manufacture of iron that they need not be repeated, but the trade of the Coppersmith is distinct from those of the workers in other metals, though they mostly consist in forming the sheet

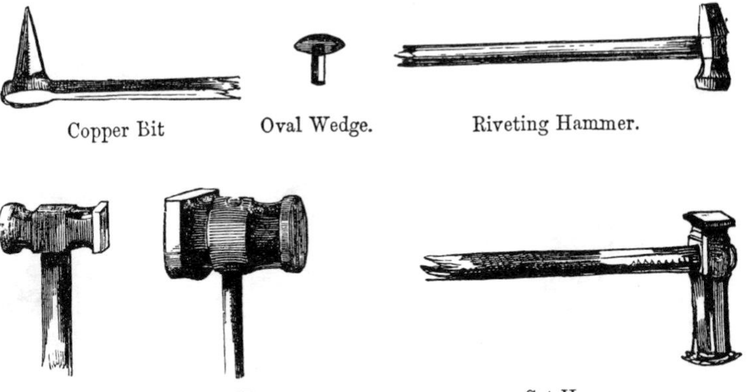

Copper Bit. Oval Wedge. Riveting Hammer.

Smoothing Hammer. Cob Hammer. Set Hammer.

copper into various shapes and utensils by means of *hammers,* of which the *smoothing hammer*, the *set hammer*, and the *riveting hammer*, explain their own uses by their names.

In manufacturing the large vessels with circular bottoms which are so frequently made of copper, the metal is first cast in a shape resembling a round spectacle glass, that is to say, a flat cake, thick in the middle and gradually diminishing in thickness towards the edge. It is then subjected to the powerful blows of a tilt hammer, the beating being principally confined to the centre. The effect of this is not only to reduce the thickness of the copper, but to

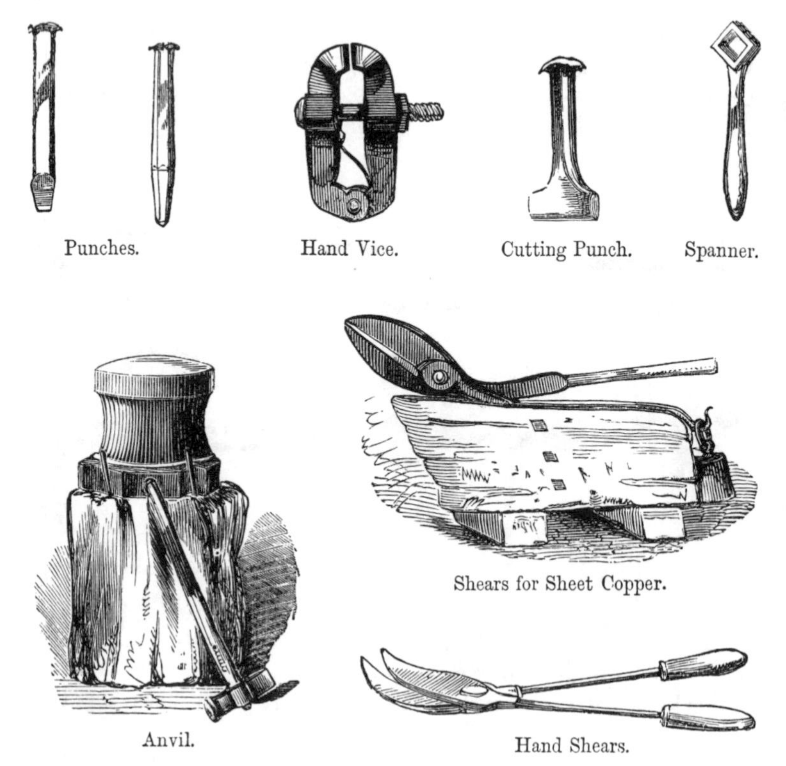

Punches. Hand Vice. Cutting Punch. Spanner.

Anvil. Shears for Sheet Copper. Hand Shears.

cause the disc to turn up at the edges and assume the form of a hollow dish.

Another operation of the Coppersmith is called *planishing*, or hammering the metal until it becomes more dense, firm, and tough; any one who looks at the surface of a large copper vessel will see the marks of the hammer by which it has been planished.

It is by the combined operation of casting, rolling, hammering, and planishing, as well as by the processes of fastening, either with *rivets* or with *solder*, that nearly all copper articles are made.

Of the tools used by the Coppersmith, beside those already mentioned, the principal are *punches* for cutting

Plumber s Iron. Iron Horse. Pouncing Block and Hammer.

or piercing holes, *shears* for cutting the sheets of metal, the *spanner* for turning heads of screws or nuts, the *anvil*, the *blocks* and *horses* for receiving the work in such a position as to operate on any part where the hammer or the punch is required, and the *iron* for soldering. There are five different modes of forming copper piping out of sheet metal; in the first the edges of the sheet, which is curved round a mandril, are made to meet without overlapping, and are joined with hard solder; in the second they overlap and are united by soft solder; in the third they overlap and are secured by rivets; in the fourth the edges are folded one over the other, and are made close and firm by hammering; and in the fifth both edges of the pipe are turned back and covered with a strip of sheet metal, the two edges of which are turned in and hammered down.

Wood Horse.

THE GUN MANUFACTURER.

THE PROOF HOUSE.

By guns, only muskets, rifles, and fowling-pieces are here meant, since the manufacture of cannons, to which the term guns is now commonly applied, is an entirely different business to that of making what are called "small arms;" the latter including, in fact, swords, pistols, bayonets, muskets, rifles, and some other implements of war. There are so many varieties of fire arms, and they are sold at prices varying so greatly, that it would be almost impossible to give any detailed description of each kind of piece. The finely-finished rifle or sporting gun, fitted with the last improvements, breech-loading or otherwise, and finished with marvellous perfection and accuracy, is worth four times as many pounds as the common muskets, made for exportation to Africa for the use of the natives, are worth shillings. We may, therefore, describe some of the ordinary processes of the gunmaker's trade, and the various improvements will then be better understood when the reader has an opportunity of seeing finished guns, in which such improvements may be pointed out to him.

The barrels of guns are either plain or twisted; twisted barrels are made of long and very narrow strips of iron; one of which, being moderately heated to increase its pliancy, is wrapped spirally round a cylindrical mandril in such a way as to form a tube, which may be slipped off the mandril at pleasure. As the rods are not usually made of sufficient length for one to form a barrel, several are usually joined end to end, those which form the breech being thicker than those at the muzzle end. By heating and hammering these pieces are welded into a continuous and very strong and tough tube.

Partially worn iron, called "scrap iron," is best for these purposes. The twisted barrels which are known as "wire twist" are formed of narrow rods of iron and steel forged together, and then rolled out to the proper thinness.

Damascus barrels are composed of similar metal, but the rods are twisted on their own centres until the fibres which they contain have from twelve to fourteen turns in an inch, by which the rods are doubled in thickness and proportionately reduced in length. Two such rods are welded together side by side, their respective twists being reversed. There are many modes of making twisted barrels, but these are the most common.

After being welded, the barrels are carefully examined, and, if needful, straightened by a few blows of the hammer. They are then bored in a machine with an angular plug of tempered steel, which is caused to revolve rapidly within the barrel, while a stream of water is directed upon the outside to check the heat caused by the tremendous friction. The outside is brought to a smooth surface either by grinding on a large grindstone, or turning in a lathe. The breech end of the barrel is bored with a screw-thread, to receive the breech plug, which closes it at that end.

Mandril, with Damascus Barrel in progress.

Tool for boring Barrel.

Testing Barrel.

The barrels are then proved by being fired at the *proof-house*, a large building where they are loaded with a charge five times as great as they will have to bear when in use. A great number of barrels are fired at once by laying them upon a strong framework of wood with their touch-holes downwards, and connected with a train of gunpowder which is ignited outside the building. A heap of sand is piled inside the building, opposite the muzzles of the barrels to receive the bullets. Those which bear this test without injury are marked as perfect.

Guns used in field sports are often made with two barrels fixed side by side upon one stock. The barrels are made separately, and each with one flat side, that they may lie close together. They are secured together by ribs running between them from end to end.

The wooden stock upon which the barrel is fixed is generally made of walnut-tree wood. It is first shaped, and afterwards shod with brass or steel; the trigger guard and other fittings are let into the wood, and every part is furnished with the proper screws and fastenings.

With regard to the manufacture of rifles and other small arms in general, the author of the present work some time ago wrote a description of a visit to the factory at Birmingham known as the Toledo Works, and it may be useful to give an extract from that account of what he saw there.

The steel from which the swords are made is supplied in long pieces somewhat tapering at the ends, and having a square portion in the middle, which, being cut through, leaves material for two blades, the bisection of the square leaving a shoulder at one end to receive the iron "tang" by which the blade is afterwards fixed into the handle. The manufacture of these blades is almost entirely effected by the forgers, who hammer them into the required shape upon the anvil, a mould running down the centre of which secures the hollow which in swords extends for about two-thirds of the length from hilt to point. In a little street of smithies the musical clink is being sounded by a score of stalwart arms, either forging the rough steel into form, or hammering the formed blade into perfect shape and symmetry, an operation which requires it to be kept at a certain heat, lest the embryo blade should be injured in the process. Once perfected as to proportion, the hardening commences, and the blade is thrust backwards and forwards into the furnace until it has acquired a proper and uniform heat, at which point it is removed and instantly plunged into cold water. This process, which has obviously suggested the Turkish bath, renders it hard indeed, but at the same time so extremely brittle that we whisperingly suggest the propriety of contracting to supply our enemies with weapons, and neglecting to carry them beyond that particular stage of preparation, when they may be snapped with the fingers. Carefully supported, however, the blade is again subjected to the fiery ordeal until it attains a slaty blue colour, and a beautiful and elastic temper, which has been partially secured by the previous hammering. By the process of forging it has become about six inches longer than the pristine steel shape, and by the tempering it has attained a springy strength which enables it to be bent in a curve sufficient to bring the hand five inches nearer to the point than when the blade is straight.

Many of the best bayonets are forged in the same way as the sword-blades, and, as in almost every manufacturing process, human intelligence has an unmistakable advantage over mere mechanical force, these possess some superior qualities. The greater number of bayonets, however, are made from a square bar of drawn steel, five inches and a half long by nine-sixteenths square. This bar is passed between a series of about sixteen pairs of rollers, which are worked by steam power, and so grooved as gradually to mould the blade to the required shape. Sixteen times the short steel bar undergoes the merciless pressure of a progressively-increased power, until its length is increased from five and a half to twenty-six inches, when some portion is cut off from the point to leave it the regulation length.

The matchets, which are made from bevel-edged steel, passed twice through the rollers, are cut into the requisite shape by means of powerful shears.

These operations are conducted in a large shed, where the rollers stand like awful combinations of infernal machines and patent mangles; where a boding and vengeful tilt-hammer, worked by steam, is tended by a man, who sits

Steam Hammer.

like a calm fate beside its crushing bulk, and supplies it with fresh victims; where the awful boom seems to shatter the very atmosphere, and deafness reigns triumphant. In obedience to a signal, however, the monster is suddenly stopped, and we are enabled to hear that the great two-pot furnace on our left is used for making the steel from those long laths of bevel-edged iron stacked against the wall; that the furnace is constructed with wide flues on each side and under the bottom, while the fire-grate occupies the centre between the two pots; that the pots themselves are some four feet deep, and two feet and a half wide, are air-tight, contain layers of charcoal and iron covered with loam sand, will remain seven days and nights in the furnace, until their contents are white hot, and that at the end of that time the iron will have been converted into steel of a slaty-blue colour. The inexorable hammer resuming its work at this point, we follow the bayonet to its completion, and once more visit the forges, to witness the "shutting on" or welding the blade to a piece of iron, which ultimately forms the socket by which the bayonet itself is fixed on to the barrel of the rifle or musket.

There is yet another operation before the blades are taken to the finishing shop, one of the most important, too, since it is no other than grinding, a process which secures an exact and uniform thickness and increases their elasticity.

We are standing at the open end of a long, vast, and gloomy shed-like building, supported by iron pillars. On each side through the entire length a series of enormous grindstones spin round amidst sand and water, and the mud from both. Seated astride the bodies of wooden horses, whose heads seem to have been transformed into these wheels, the grinders seize upon the blades, and each fearless rider rising in his stirrups, or, what looks much the same, standing tiptoe till he no longer touches his saddle, throws himself forward, and presses the sword, matchet, or bayonet on the wheel, at the same time guiding it deftly with its left hand, till its whole surface has been smoothly ground.

Along the whole line of whirling stones fly the lurid red sparks, and the grinders, with squared elbows, seem to curb the struggling and impetuous wheels.

After polishing, which is completed by wooden wheels bearing a coating of leather covered with emery, the swords and matchets go to receive handles, and the bayonets locking rings. The handles of swords are made of walnut-wood covered with the skin of the dogfish, while the hilt and guard are formed from a plain flat sheet of steel, in shape not unlike one side of a pair of bellows.

The solid socket of the bayonet is hammered into form, and afterwards stamped into shape with the rim complete, from which process it is conveyed to a shop where it is drilled by steam power. It then only remains to secure a smooth surface, by means of a revolving barrel, containing an instrument with a number of flanged blades, against which the socket of the bayonet is pressed. It is not a little remarkable to see the solid steel pared and shaved like wax, and no less wonderful to notice the simple machinery by which it is accomplished. The locking rings are stamped out by a lever and die, pierced by a punch, and afterwards "bored," "faced," and their shapes secured by a triple circular saw, worked by a lathe.

The most important manufacture in the Toledo Works, however, is assuredly rifles, and, with the intention of following it through its principal processes, we return to the vicinity of the still inveterate hammer, where we are shown a rudimentary barrel in the shape of a slab of best wrought iron, twelve inches long, and weighing nine pounds and a quarter. This uninviting slab is heated in a furnace, and roughly bent into the tubular shape by means of our enemy the tilt-hammer, after which it is once more placed in a furnace of an enormously high temperature, with a small trap-opening. When sufficiently heated, the short rudimentary tube is taken out on a long round iron rod, fitted with a hand-guard, and looking like a huge burlesque rapier. This rod approximates to the size of the intended bore of the barrel, and is inserted, with the rough tube upon it, between two steam rollers, each of which is furnished

Mandril, with Common Musket Barrel in progress.

Welding by Hand.

Forging Hammer.

with a series of corresponding grooves or cuts. The barrel, which is taken up at one end by a rod, is placed between the first pair of grooves, and, as the rollers revolve, is drawn out at the other side, a long, hollow, welded tube. This much more graceful and better formed tube is then consigned to another rod of smaller diameter, and to a corresponding pair of grooves; until, after the eighth repetition of the same process, the barrel has attained its proper dimensions. The next operation, which is called "lumping," consists of welding a piece of wrought iron on to the breech end of the barrel, for the purpose of forming the percussion-lump, and is succeeded by "rough-boring." This is accomplished by a long, sharp-ended bit, which, being placed in the end of the barrel, revolves at the rate of, perhaps, a thousand turns a minute by means of a pulley and fly-wheel, while the barrel is pushed on by a lever, and kept cool by means of water thrown upon its surface.

The "setting" of the barrel is next effected by means of hammer and anvil, the setting meaning simply rectifying any bend which it may have received during the previous operation. We are not a little interested in the setting, since the first intimation of it on entering the shop is the sudden discovery of a number of workmen gazing resolutely at an opposite window, through what look like attenuated telescopes. They are engaged, however, in one of the processes which require the greatest experience, as each of them is expected to detect the most trifling bias in the barrel. The "spilling-up," or cutting the inside of the barrel to the proper bore, is similar to the rough-boring, except that only one edge of the bit is allowed to operate, the others being sheathed by a half cylinder of wood, called a *spill*; this ensures a smooth surface, and prepares for the "fine-boring," which is six times repeated, the final surface being insured by keeping one edge of the bit perfectly smooth, by which means the particles of steel drop in a fine and almost soft powder.

The outside of the barrel is next turned in a long lathe, which not only reduces the roughness, but, by a beautiful arrangement of cutting tools, gives it the required substance or "pattern," for a light or heavy rifle.

The grinding of the barrels is effected by means of stones, larger than those used for the sword blades, but in a similar manner, and is preliminary to "filing," which carries the barrel to the shop, where it is prepared for the lock.

Grinding Gun Barrels.

These preparations consist of "chambering," or making the chamber which holds the pin; "breeching," or cutting the worm intended for the breech-pin, that helps to hold the barrel to the stock by means of a breech-nail; cutting out the little slice into which the "sight" is to be dovetailed;

machining the lump; filing the tail-pin, and making the square lump the proper shape for receiving the lock and stock.

We are not a little surprised to learn that every part of the lock is finished by hand, the cock being cut with a die worked by a heavy weight, and the smaller pieces being wrought with forge, hammer, and file.

The great art in lock-making is to obtain a perfect spring, and those properly tempered are so elastic that, although when fitted in the lock, the two sides are so close as almost to touch, they will, when released, spread to two inches below the edge of the lock-plate. The lock and

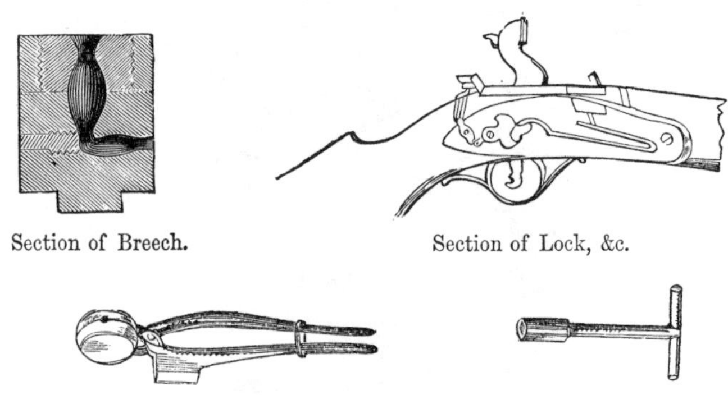

Section of Breech. Section of Lock, &c.

Bullet Mould. Nipple Key.

barrel are now ready for the stock, which awaits them in another shop, where it has been sawn out of walnut wood, and finished by carpenters' tools. The barrel let into its groove, and the lock properly in its place, the stock is more perfectly shaped and rounded before "screwing together," or the addition of the different parts of the "furniture," heel-plate, trigger-plate and guard, trigger, nose-cap, rod, and bayonet.

We are now told that the rifle is "finished," by which, understanding *completion*, we are not quite prepared to learn that it is to be taken to pieces.

We suddenly remember, however, that it is not yet a rifle at all, inasmuch as it has not been rifled. Everything is made perfect before this delicate operation is attempted, in order that no injury may be sustained by the barrel when the complete rifle is again put together. The process of rifling is similar to that of boring, except that a spiral cutter is substituted for the bit. Previous to the reunion of the barrel, the whole work is polished, and the stock stained and finished ready for completion.

"Finishing."

The pistol barrels undergo the same processes as that of the rifle, except that, after being drilled, they are planed, by machines which carry them along a sort of bed under tools that cut them perfectly smooth, and accurately shape the octagonal barrels. These chisels move by means of screws over the entire surface as it is drawn backwards and forwards on the slide.

The revolver chambers are drilled out of solid iron, by a drilling machine or lathe, with a centrebit and an eccentric motion, which causes each barrel of the chamber-nest to become the centre in succession; while, by means of a slide, the motion can be made to suit either a large or small chamber. The recesses communicating with the lock and trigger are cut by reversing the chamber in the eccentric "chuck," and using a different cutting-tool, while another alteration effects the drilling of the nipple holes.

THE END